Microbiology Experiments:
A Health Science Perspective

John Kleyn

Mary Bicknell
University of Washington

Marie Gilstrap

WCB
Wm. C. Brown Publishers

Dubuque, IA   Bogota   Boston   Buenos Aires   Caracas   Chicago
Guilford, CT   London   Madrid   Mexico City   Sydney   Toronto

Book Team

Editor *Elizabeth M. Sievers*
Developmental Editor *Robin P. Steffek*
Production Editor *Marla K. Irion*
Designer *Lu Ann Schrandt*
Art Editor *Kathleen Huinker Timp*
Photo Editor *Lori Hancock*
Permissions Coordinator *Mavis M. Oeth*

WCB **Wm. C. Brown Publishers**

President and Chief Executive Officer *Beverly Kolz*
Vice President, Publisher *Kevin Kane*
Vice President, Director of Sales and Marketing *Virginia S. Moffat*
Vice President, Director of Production *Colleen A. Yonda*
National Sales Manager *Douglas J. DiNardo*
Marketing Manager *Patrick E. Reidy*
Advertising Manager *Janelle Keeffer*
Production Editorial Manager *Renée Menne*
Publishing Services Manager *Karen J. Slaght*
Royalty/Permissions Manager *Connie Allendorf*

Times Mirror
Higher Education Group

President and Chief Executive Officer *G. Franklin Lewis*
Senior Vice President, Operations *James H. Higby*
Corporate Senior Vice President, President of WCB Manufacturing *Roger Meyer*
Corporate Senior Vice President and Chief Financial Officer *Robert Chesterman*

Copyedited by Sarah Aldridge

Cover Photos: Scanning electron micrograph of the fungus *Aspergillus nidulans*: © E. Gueho/CNRI/SPL/Photo Researchers, Inc. Inset: Scanning electron micrograph of yeast on the human tongue: © Veronika Burmeister/ Visuals Unlimited. Part opening (bright-field micrograph of *Aspergillus*): Courtesy of Dr. T. Fwitsche, University of Washington.

Illustrations by Mark Nero: 1.1, 2.2, 2.4, 2.5, 2.6, 3.2a–c, 4.1, 4.2, 4.3, 5.1, 5.2, 5.3, 5.4, 6.1, 6.2, 6.3, 7.1, 8.1, 8.2, 8.3, 9.1, 9.2, 9.3, I.3.1, I.5.1, 10.1, 10.2, 11.1, 12.1, 12.2, 13.1, 13.2, 14.1a–b, 14.2, 15.1, 15.2, 15.3, 16.1, 16.2, 17.1a–b, 17.2, 18.1, 18.2, 19.1a–d, 20.1, 20.2, 20.3, 20.4, 21.1, 21.3b, 21.5, 22.1, 22.3, 22.4, 23.1, 24.1, 24.2, 24.3, 25.1, 27.1, 27.2, 29.1, 29.2, 29.3, 30.2, 31.1, 31.2, 31.3, 32.1, 32.2, 32.3, 32.4, 34.1, 34.4, 34.5 and text art pages 34, 70, 187, 280.

Some of the laboratory experiments included in this text may be hazardous if materials are handled improperly or if procedures are conducted incorrectly. Safety precautions are necessary when you are working with microorganisms, chemicals, glass test tubes, hot water baths, sharp instruments, and the like, or for any procedures that generally require caution. Your school may have set regulations regarding safety procedures that your instructor will explain to you. Should you have any problems with materials or procedures, please ask your instructor for help.

Printed in the United States of America by Times Mirrror Higher Education Group, Inc., 2460 Kerper Boulevard, Dubuque, IA 52001

10 9 8 7 6 5 4 3 2 1

Dedication

To our mothers who tirelessly encouraged us.

To our spouses who supported us with their patience, suggestions, and good humor.

To our students who caught the excitement of the microbial world.

Contents

PART FIVE
Public Health and the Role of Microbiology in the Prevention and Control of Communicable Diseases   283

APPENDICES

Preface

To the Student

A microbiology laboratory is valuable because it actually gives you a chance to see and study microorganisms firsthand. In addition, it also provides you with the opportunity to learn the special techniques used to study and identify these organisms. The ability to make observations, record data, and analyze results is useful throughout life.

It is very important to read the scheduled exercises before coming to class, so that class time can be used efficiently. It is helpful to ask yourself the purpose of each step as you are reading and carrying out the steps of the experiment. Sometimes it will be necessary to read an exercise several times before it makes sense.

Conducting experiments in microbiology laboratories is particularly gratifying because the results can be seen in a day or two (as opposed, for instance, to plant genetics laboratories). Opening the incubator door to see how your cultures have grown and how the experiment has turned out is a pleasurable moment. We hope you will enjoy your experience with microorganisms as well as acquire skills and understanding that will be valuable in the future.

To the Instructor

The manual includes a wide range of exercises—some more difficult and time-consuming than others. Usually more than one exercise can be done in a two-hour laboratory period. In these classes, students can actually see the applications of the principles they have learned in the lectures and text. We have tried to integrate the manual with the text *Microbiology: A Human Perspective* by Nester et al.

The exercises were chosen to give students an opportunity to learn new techniques and to expose them to a variety of experiences and observations. It was not assumed that the school or department had a large budget, thus exercises have been written to use as little expensive media and equipment as possible. The manual contains more exercises than can be done in one course so that instructors will have an opportunity to select the appropriate exercises for their particular students and class. We hope that the instructors find these laboratories an enjoyable component of teaching microbiology.

Acknowledgments

We would like to acknowledge the contributions of the lecturers in the Department of Microbiology at the University of Washington who have thoughtfully honed laboratory exercises over the years until they really work. These include Dorothy Cramer, Carol Laxson, Mona Memmer, Janis Fulton, and Denise Anderson. Special thanks to Dale Parkhurst for his expert knowledge of media.

We also want to thank Eugene and Martha Nester and Evans Roberts for their text *Microbiology: A Human Perspective*. This text was the source of much of the basic conceptual material and figures for our laboratory manual.

Additional thanks to Abbott Laboratories, Becton Dickinson, Difco Laboratories, and Meridian Laboratories for their generous offers to make diagnostic kits available for some exercises.

Reviewers

Helen M. Finley
Milwaukee Area Technical College

Marianne Crocker
Ozark Technical Community College

Steven Woeste
Scholl College of Podiatric Medicine

Laboratory Safety

To be read by the student before beginning any laboratory work.

1. Do not eat, drink, smoke, or store food in the laboratory. Avoid all finger-to-mouth contact.
2. Never pipette by mouth because of the danger of ingesting microorganisms or toxic chemicals.
3. Wear a laboratory coat while in the laboratory. Remove it before leaving the room and store it in the laboratory until the end of the course.
4. Wipe down the bench surface with disinfectant before and after each laboratory period.
5. Tie long hair back to prevent it from catching fire in the Bunsen burner or contaminating cultures.
6. Keep the workbench clear of any unnecessary books or other items. Do not work on top of the manual because if spills occur, it cannot be disinfected easily.
7. Be careful with the Bunsen burner. Make sure that paper, alcohol, the gas hose, and your microscope are not close to the flame.
8. All contaminated material and cultures must be placed in the proper containers for autoclaving before disposal or washing.
9. Avoid creating aerosols by gently mixing cultures. Clean off the loop in a sand jar before flaming in the Bunsen burner.
10. If a culture is dropped and broken, notify the instructor. Cover the contaminated area with a paper towel and pour disinfectant over the material. After ten minutes, put the material in a broken glass container to be autoclaved.
11. Carefully follow the techniques of handling cultures as demonstrated by the instructor.
12. When the laboratory is in session, the doors and windows should be shut. A sign should be posted on the door indicating that it is a microbiology laboratory.
13. Be sure you know the location of fire extinguishers, eyewash apparatus, and other safety equipment.
14. Wash your hands with soap and water after any possible contamination and at the end of the laboratory period.

Correlation Table

Correlation of Exercises in *Microbiology Experiments: A Health Science Perspective* to Chapters in Some Major Texts

Exercise	Nester/Roberts/ Nester *Microbiology* 1/e (WCB)	Talaro/Talaro *Foundations in Microbiology* 1/e (WCB)	Tortora/Funke/ Case *Microbiology* 5/e (Benjamin/ Cummings)	Black *Microbiology* 2/e (Prentice-Hall)	Pelczar/Chan/ Krieg *Microbiology* 1/e (McGraw-Hill)	Alcamo *Fundamentals of Microbiology* 4/e (Benjamin/ Cummings)
1. Ubiquity of Microorganisms	1/3	1/3	1/3	1/3	2/3	1/3
2. Bright-field Light Microscopy, Including History and Working Principles	3	1	3	3	3	3
3. Microscopic (Bright-field and Dark-field) Determination of Cell Motility, Form, and Viability Using Wet Mount and Hanging Drop Preparations	3	1	3	3	3	3
4. Simple Stains: Positive and Negative Stains	3	1	3	3	3	3
5. Multiple and Differential Stains	3	1	3	3	3	3
6. Pure Culture and Aseptic Technique	4/9	6/9	6/7	6/13	6/7/8	4/21/22
7. Defined, Undefined, Selective, and Differential Media	4	6	6	6	5/6	4
8. Quantification of Microorganisms	4	6	6	6	6	4
9. Aerobic and Anaerobic Growth	4	6	6	6	6	4
10. The Effect of Incubation Temperature on Generation Time	4	6	6	6	6	4

Continued

Correlation of Exercises in *Microbiology Experiments: A Health Science Perspective* to Chapters in Some Major Texts—*Continued*

Exercise	Nester/Roberts/ Nester *Microbiology* 1/e (WCB)	Talaro/Talaro *Foundations in Microbiology* 1/e (WCB)	Tortora/Funke/ Case *Microbiology* 5/e (Benjamin/ Cummings)	Black *Microbiology* 2/e (Prentice-Hall)	Pelczar/Chan/ Krieg *Microbiology* 1/e (McGraw-Hill)	Alcamo *Fundamentals of Microbiology* 4/e (Benjamin/ Cummings)
11. Moist and Dry Heat Sterilization: Thermal Death Point and Thermal Death Time	4/9	6/9	7/8	6/13	6/7	4/21
12. Control of Microbial Growth with Ultraviolet Light	4/9	6/9	7/8	6/13	6/7	4/21
13. Osmotic Pressure, and the Rate and Amount of Microbial Growth	4/9	6/9	7/8	6/13	6/7	4/21
14. Antiseptics and Antibiotics	9	9/10	7/20	13/14	8/21	22/23
15. Selection of Bacterial Mutants Resistant to Antibiotics	7	8	8	7	13	6
16. Transformation: A Form of Genetic Recombination	7	8	8	7	13	6
17. Bacterial Conjugation	7	8	8	7	13	6
18. Gene Regulation: Induction and Catabolite Repression	6/7	8	8	7	13	6
19. Nonfilamentous Fungi: Yeasts and Yeastlike Fungi	12	4	12	12	10	14
20. Filamentous Fungi: Particularly Microscopic Forms Such as Molds	12	4	12	12	10	14
21. Parasitology: Protozoa and Helminths	12	4	12	12	10	15/16
22. Prokaryotic Viruses: Bacteriophage Isolation and Titering	13	5	13	11	15/16	11
23. Normal Skin Flora	15/19	11	14/21	15/20	17/27	10/17
24. Respiratory Microorganisms	20/21	15/16/17	24	21	24	7

Continued

Correlation of Exercises in *Microbiology Experiments: A Health Science Perspective* to Chapters in Some Major Texts—*Continued*

Exercise	Nester/Roberts/ Nester *Microbiology* 1/e (WCB)	Talaro/Talaro *Foundations in Microbiology* 1/e (WCB)	Tortora/Funke/ Case *Microbiology* 5/e (Benjamin/ Cummings)	Black *Microbiology* 2/e (Prentice-Hall)	Pelczar/Chan/ Krieg *Microbiology* 1/e (McGraw-Hill)	Alcamo *Fundamentals of Microbiology* 4/e (Benjamin/ Cummings)
25. Identification of Enteric Gram-negative Rods	10/11	3/16	10/11	9/10	3/9	4/8
26. Clinical Unknown Identification	10/19-26	14	10	6	9	8
27. Differential White Blood Cell Stains	15/16	12/13	16/17	15/17	19	18/19
28. Lysozyme, an Enzymatic Form of Natural Resistance	16	12	16/17	17/18	18/19	18/19
29. Traditional Agglutination Reactions Employing Microbial and Nonmicrobial Antigens	17	13	18	19	20	19
30. Lancefield Grouping of Pathogenic Streptococci with a Latex Slide Agglutination Test	17/20	13/15	18/24	19/21	20/24	7/19
31. Use of an Enzyme-Linked Immunoassay (ELISA) Test for Group A *Streptococcus* Identification	17	13	18	19	20	19
32. An Ouchterlony Double Immunodiffusion Test for *Candida* Identification	17	13	18	19	20	19
33. Epidemiology: A *Staphylococcus* Carrier Study	18	11	14	16	22	17/18
34. Bacteriological Examination of Water: Multiple-Tube Fermentation and Membrane Filter Techniques	31	22	27	26	29	26

PART ONE

Basic Microbiology

Introduction to Microbiology

When you take a microbiology class, you have an opportunity to explore an extremely small biological world that exists unseen in our own ordinary world. Fortunately, we were born after the microscope was perfected so we can see these extremely small organisms.

A few of these many and varied organisms are pathogens (capable of causing disease). Special techniques have been developed to isolate and identify them as well as to control or prevent their growth. The exercises in this manual will emphasize medical applications. The goal is to teach you basic techniques and concepts that will be useful to you now or can be used as a foundation for additional courses. In addition, these exercises are also designed to help you understand basic biological concepts that are the foundation for applications in all fields.

As you study microbiology, it is also important to appreciate the essential contributions of microorganisms as well as their ability to cause disease. Most organisms play indispensable roles in breaking down dead plant and animal material into basic substances that can be used by other growing plants and animals. Photosynthetic bacteria are an important source of the earth's supply of oxygen. Microorganisms also make major contributions in the fields of antibiotic production, food preservation, and, more recently, recombinant DNA technology. The principles and techniques demonstrated here can be applied to these fields as well as to medical technology, nursing, or patient care. This course is an introduction to the microbial world, and we hope you will find it useful and interesting.

Note: The use of pathogenic organisms has been avoided whenever possible, and nonpathogens have been used to illustrate the kinds of tests and procedures that are actually carried out in clinical laboratories. In some cases, however, it is difficult to find a substitute and organisms of low pathogenicity are used. These exercises will have an additional safety precaution.

Nothing is more satisfying to the mind than to be able to follow a discovery from its very origin to its latest developments.

Louis Pasteur

EXERCISE 1

Ubiquity of Microorganisms

Getting Started

Microorganisms are everywhere—in the air, soil, and water; on plant and rock surfaces; and even in such unlikely places as Yellowstone hot springs and Antarctic ice. Millions of microorganisms are also found living with animals—for example, the mouth, the skin, the intestine all support huge populations of bacteria. In fact, the interior of healthy plant and animal tissues is one of the few places free of microorganisms. In this exercise, you will sample material from the surroundings and your body to determine what organisms are present that will grow on laboratory media.

An important point to remember as you try to grow organisms, is that there is no one condition or **medium** that will permit the growth of all microorganisms. The trypticase soy agar used in this exercise is a rich medium (a digest of meat and soy products, similar to a beef and vegetable broth) and will support the growth of many diverse organisms, but bacteria growing in a freshwater lake that is very low in organic compounds would find it too rich (similar to a goldfish in vegetable soup). However, organisms that are accustomed to living in our nutrient-rich throat might find the same medium lacking necessary substances they require.

Temperature is also important. Organisms associated with warm-blooded animals usually prefer temperatures close to 37° C, which is approximately the body temperature of most animals. Soil organisms generally prefer a cooler temperature of 30° C. Organisms growing on glaciers would find room temperature (about 25° C) much too warm and would probably grow better in the refrigerator.

Microorganisms also need the correct atmosphere. Many bacteria require oxygen, while other organisms find it extremely toxic and will only grow in the absence of air. Therefore, the organisms you see growing on the plates may be only a small sample of the organisms originally present.

Definitions

Agar. A carbohydrate derived from seaweed used to solidify a liquid medium.

Colony. A visible population of microorganisms growing on a solid medium.

Inoculate. To transfer organisms to a medium to initiate growth.

Media (medium, singular). The substance used to support the growth of microorganisms.

Pathogen. An organism capable of causing disease.

Sterile. The absence of either viable microorganisms or viruses capable of reproduction.

Ubiquity. The existence of something everywhere at the same time.

Objectives

1. To demonstrate that organisms are ubiquitous.
2. To demonstrate how organisms are grown on laboratory culture media.

References

Nester et al. *Microbiology: A human perspective.* Chapter 4, 89.

Materials

Per team of two (or each individual, depending on amount of plates available)

Trypticase soy agar (TSA) plates, 2

Sterile swabs as needed

Sterile water (about 1 ml/tube) as needed

Waterproof marking pen or wax pencil

Figure 1.1 Plates labeled on the bottom for ubiquity exercise.

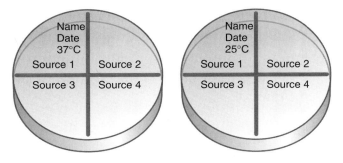

Procedure

First Session

1. Each pair of two students should obtain two Petri plates of trypticase soy agar. Notice that the lid of a Petri plate fits loosely over the bottom half.
2. Label the plates with your name and date using a wax pencil or waterproof marker. Always label the bottom of the plate because sometimes you may be examining many plates at the same time and it is easy to switch the lids.
3. Divide each plate in quarters with two lines on the back of the Petri plate. Label one plate 37° C and the other 25° C (figure 1.1).
4. Inoculate the 37° C plate with samples from your body. For example, moisten a sterile swab with saliva and rub it on one of the quadrants. Try touching your fingers to the agar before and after washing or place a hair on the plate. Try whatever interests you. (Be sure to place all used swabs into an autoclave container or bucket of disinfectant after use.)
5. Inoculate the plate labeled 25° C (room temperature) with samples from the room. It is easier to pick up a sample if the swab is moistened in sterile water first. Sterile water is used so that there will be no living organisms in the water to contaminate your results. Try sampling the bottom of your shoe or some dust, or press a coin or other objects lightly on the agar. Be sure to label each quadrant so that you will know what you used as inoculum.
6. Incubate the plates at the temperature written on the plate. Place the plates in the incubator or basket upside down. This is important because it prevents condensation from forming on the lid and dripping on the agar below. The added moisture would permit colonies of bacteria to run together.

Second Session

Handle all plates with colonies as if they were potential pathogens. Follow your instructor's directions carefully.

Note: For best results, the plates incubated at 37° C should be observed after 2 days, but the plates at room temperature will be more interesting at about 5–7 days. If possible, place the 37° C plates either in the refrigerator or at room temperature after 2 days so that all the plates can be observed at the same time.

1. Examine the plates you prepared in the first session and record your observations on the report sheet for this exercise. There will be basically two kinds of colonies: fungi (molds) and bacteria. Mold colonies are usually large and fluffy, the type found on spoiled bread. Bacterial colonies are usually soft and glistening, and tend to be cream colored or yellow. Compare your colonies with color plates 1 and 2.
2. When describing the colonies include:
   a. relative size as compared to other colonies
   b. shape (round or irregular)
   c. color
   d. surface (shiny or dull)
   e. consistency (dry, moist, or mucoid)
   f. elevation (flat, craterlike, or conical)
   There are standardized words to describe colonies, but for this exercise use your own descriptive terms.
3. There may be surprising results. If you pressed your fingers to the agar before and after washing, you may find more organisms on the plate after you washed your hands. The explanation is that your skin has a normal flora (organisms that are always found growing on your skin). When you wash your hands, you wash off the organisms you have picked

up from your surroundings as well as a few layers of skin. This exposes more of your normal flora; therefore, you may see different colonies of bacteria before you wash your hands than afterward. Your flora is important in preventing undesirable organisms from growing on your skin. Hand washing is an excellent method for removing pathogens that are not part of your normal flora.

4. (Optional) If desired, use these plates to practice making simple stains or Gram stains in exercises 5 and 6.

Note: In some labs, plates with molds are opened as little as possible and immediately discarded in an autoclave container to prevent contaminating the lab with mold spores.

5. Follow the instructor's directions for discarding plates. All agar plates are autoclaved before washing or discarding them in the municipal garbage system.

NOTES:

EXERCISE

1

Laboratory Report: Ubiquity of Microorganisms

Results

Room Temperature (about 25° C) Plate

| | Plate Quadrant | | | |
	1	2	3	4
Source				
Colony appearance				

37° C Plate

| | Plate Quadrant | | | |
	1	2	3	4
Source				
Colony appearance				

Questions

1. Give three reasons that all the organisms you placed on the TS agar plates might not grow.

2. Why were some agar plates incubated at 37° C and others at room temperature?

3. Why do you invert agar plates when placing them in the incubator?

4. Name one area that might be free of microorganisms.

EXERCISE 2

Bright-field Light Microscopy, Including History and Working Principles

Getting Started

Microbiology is the study of living organisms too small to be seen with the naked eye. An optical instrument, the microscope, allows you to magnify microbial cells sufficiently for visualization. The objectives of this exercise are to inform you about: (1) some pertinent principles of microscopy; and (2) the practical use, including instruction and care, of the bright-field light microscope.

Historical

Anton van Leeuwenhoek (1632–1723), a Dutch linen draper and haberdasher, recorded the first observations of living microorganisms using a home-made microscope containing a single glass lens (figure 2.1) powerful enough to enable him to see what he described as little "animalcules" (now known as bacteria) in scrapings from his teeth, and larger "animalcules" (now known as protozoa and algae) in droplets of pond water and hay infusions.

Figure 2.1 Model of a van Leeuwenhoek microscope. The original was made in 1673 and could magnify the object being viewed almost 300 times. The object being viewed is brought into focus with the adjusting screws. This replica was made according to the directions given in the *American Biology Teacher* 30:537,1958. Note its small size. Photograph Courtesy of J.P. Dalmasso

Lens

Object being viewed

Viewing side

Adjusting screws

1 inch

A single lens microscope such as van Leeuwenhoek's had many disadvantages. Optically, they included production of distortion with increasing magnifying powers and a decrease in focal length (the distance between the specimen when in focus and the tip of the lens). Thus, when using a single lens with an increased magnifying power, van Leeuwenhoek had to practically push his eye into the lens in order to see anything.

Interestingly enough at an earlier date, 1590, Zacharias Janssen, a Dutch spectacle maker, and his son invented a microscope containing an ocular eyepiece that allowed the viewer to examine the image made by a simple lens from a reasonable distance. Such a microscope with more than one lens is called a **compound microscope.** A much advanced modification of the Janssen's compound microscope is shown in figure 2.2. Compare figure 2.2 with a contemporary teaching microscope, a bright-field compound microscope (figure 2.3).

Another giant in the early development of light microscopy was a German physicist, Ernst Abbe, who in 1883 helped to improve immensely the Janssen-type light microscopes by the addition of a third lens, the **condenser lens** (see figure 2.3 for location below microscope stage). By focusing this lens up or down, it was possible to concentrate (intensify) the light emanating from the light source on the bottom side of the specimen. The specimen, in turn, was located on the top surface of the specimen stage. He also developed the technique of using lens immersion oil in place of water as a medium for transmission of light rays from the specimen to the lens of the **oil-immersion objective.** Oil with a density more akin to the microscope lens than that of water helped decrease the loss of transmitted light, which, in turn, increased image clarity. Finally, Abbe developed improved microscope objective lenses that were able to reduce both **chromatic** and **spherical** lens aberrations. His objectives included the addition of a concave (glass bent inward like a dish) lens to the basic convex (glass bent outward) lens in order to diverge the peripheral rays of light slightly to form an almost flat image. The earlier simple convex lenses produced distorted image shapes due to spherical lens aberrations and distorted image colors due to chromatic lens aberrations.

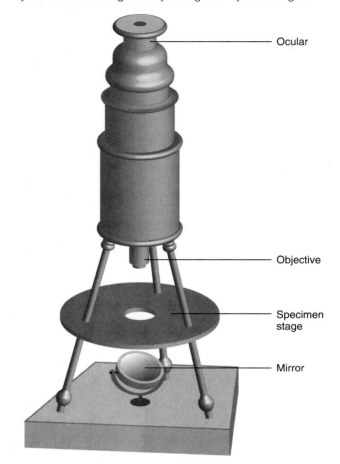

Figure 2.2 An early 1700s compound microscope with an ocular eyepiece. The specimen stage with the hole in the middle was where the specimen, to be observed, was placed. Focusing was by sliding the inner tube of the fixed cylinder. Illuminating was by sunlight or by candlelight.

Ocular

Objective

Specimen stage

Mirror

Spherical Lens Aberrations These occur because light rays passing through the edge of a convex lens are bent more than light rays passing through the center. The simplest correction is the placement of a diaphragm below the lens so that only the center of the lens is used (locate **iris diaphragm** in figure 2.3). Such aberrations can also be corrected by grinding the lenses in special ways.

Chromatic Lens Aberrations These occur because light is refracted (bent) as well as dispersed by a lens. The blue components of light are bent more than the red components. Consequently, the blue light, which is bent the most, travels a shorter distance through the lens before converging to form a

Figure 2.3 Modern bright-field compound microscope. Courtesy of Carl Zeiss, Inc.

Eyepiece—a magnifying lens, usually about 10X

Objective lens—a number of objective lenses that provide different magnifications (The total magnification of the microscope equals the *product* of the magnification of the ocular and objective lenses.)

Specimen stage—the platform on which the slide is placed

Iris diaphragm—controls the amount of light that enters the objective lens

Coarse adjustment knob

Fine adjustment knob

Condenser—focuses the light

Base with illuminating light source

blue image. The red components, which are bent the least, travel a longer distance before converging to form a red image. When these two images are seen in front view, the central area, in which all the colors are superimposed, maintains a white appearance. The red image, which is larger than the blue image, projects beyond the central area, forming red edges outside of the central white image. Correction of a chromatic aberration is much more difficult than correction of a spherical aberration since dispersion differs in different kinds of glass. Objective lenses free of spherical and chromatic aberrations, known as **apochromatic objectives,** are now available but are also considerably more expensive than **achromatic objectives.**

Microscope Image Magnification with a Modern Compound Microscope

Figure 2.4 shows a diagram of this system. The lens at the bottom end of the body tube **(objective lens)** is a powerful magnifying glass containing two or more lenses, and produces an enlarged image of the object O at I_1. The magnified image I_1 is further magnified with the ocular at the top of the body tube. The first ocular lens projects a magnified image of I_1 at I_2, and the second ocular lens produces a small image of I_2 at I_3, which is called

Ramsden's disk. This image is quite small (approx. 1 mm), and can be seen by lowering a piece of translucent paper just above the ocular until Ramsden's disk is in focus. The human eye is then able to transform the disk into the illusion of a magnified image of the object. In this instance, the lens of the eye projects an image of Ramsden's disk over the entire retina surface at point I_4. The distance between the outer top lens ocular surface and the point of focus of Ramsden's disk is known as the **eye relief.** For most oculars, this distance is so short that those who wear glasses need to remove them in order to focus Ramsden's disk on the cornea. Some manufacturers now offer oculars with long eye relief.

Some Working Principles of Bright-field Microscopy

Subjects to be covered in this section include microscope objectives, magnification and resolution, and illumination.

Microscope Objectives—The Heart of the Microscope

All other parts of the microscope are involved in helping the objective attain a noteworthy image. Such an image is not necessarily the *largest* but the *clearest.* A clear image helps achieve a better

Figure 2.4 Optical physics of image produced by a compound microscope (see text for description).

Eye

I_4 Image of I_3

I_3 Image of I_2

I_2 Image of I_1

Ocular

I_1 Image of O

Objective

O object

length (λ) of light used for illumination, as well as on the numerical apertures (NA) of the objective and condenser lens systems. These are related by the equation:

$$\text{resolving power (r)} = \frac{\lambda}{\text{NA}_{\text{obj}} + \text{NA}_{\text{cond}}}$$

Examining the equation, we can see that the shorter the wavelength employed, the greater the resolving power; conversely, the larger the numerical aperture employed, the greater the resolving power. We might say that blue light affords a better resolving power than red light because its wavelength is considerably shorter. However, because the range of the visible light spectrum is rather narrow, increasing the resolution by decreasing the wavelength is of limited use. Thus, the greatest boost to the resolving power is attained by increasing the numerical aperture of the condenser and objective lens systems.

By definition, the numerical aperture = n sin theta. The refractive index, n, refers to the medium employed between the objective lens and the specimen; e.g., with the dry low and high power objectives the medium is air, which has a refractive index of 1, whereas with the oil-immersion objective the medium is oil, which has a refractive index of 1.56. Sin theta is the maximum angle formed by the light rays coming from the condenser and passing through the specimen into the front lens of the objective.

Ideally, the numerical aperture of the condenser should be as large as the numerical aperture of the objective, or the latter is reduced, resulting in reduced resolution. Practically, however, the condenser numerical aperture is somewhat less because the condenser iris has to be closed partially in order to avoid glare. It is also important to remember that the numerical aperture of the oil-immersion objective depends upon the use of a dispersing medium with a refractive index greater than that of air ($n = 1$). This is achieved by using oil ($n = 1.56$), which must be in contact with both the condenser lens (below the slide) and the objective lens (above the slide).

Note: Oil should not be placed on the surface of the condenser lens unless your microscope contains an oil-immersion type condenser lens and your instructor authorizes its use.

understanding of specimen structure. Size alone does not help achieve this end. The ability of the microscope to reveal specimen structure is called **resolution,** whereas the ability of the microscope to increase specimen size is called **magnification.**

Resolving power is also defined as the ability of an objective to distinguish two nearby points as distinct and separate. The maximum resolving power of the human eye when reading is 0.1 mm (100 micrometers). We now know that the maximum resolving power of the light microscope is approximately 0.2 μm or 500× better than the human eye, and that it is dependent on the wave-

Figure 2.5 Relationship between working distance of objective lens and diameter of opening of the condenser iris diaphragm. The larger the working distance, the smaller the opening of the iris diaphragm.

Figure 2.6 The diagram shows that light refracts (bends) more when it passes through air (refractive index $n=1$) than when it passes through oil ($n=1.6$). Thus, by first passing the light from the light source through oil, light energy is conserved. This conservation in light energy helps to increase the resolving power of the oil-immersion objective, which also has a refractive index greater than 1 ($n = 1.25$ to 1.35).

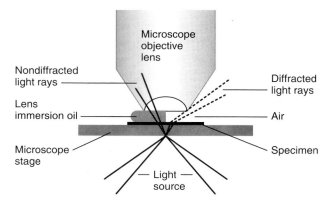

When immersion oil is used on only one side of the slide, the maximum numerical aperture of the oil-immersion objective would be 1—the same as the refractive index of air. It should be noted that immersion oil should only be used if the condenser contains an oil-immersion lens.

Microscopes for bacteriological use are usually equipped with three objectives: 16 mm low power (10×), 4 mm high dry power (40 to 45×), and 1.8 mm oil immersion (100×). The desired objective is rotated into place by means of a revolving nosepiece (see figure 2.3). The millimeter number (16, 4, 1.8) refers to the **focal length** of each objective. By definition, the focal length is the distance from the principal point of focus of the objective lens to the principal point of focus of the specimen. Practically speaking, one can say that the shorter the focal length of the objective, the shorter the **working distance** (that is, the distance between the lens and the specimen) and the larger the opening of the condenser iris diaphragm required for proper illumination (figure 2.5).

The power of magnification of the three objectives is indicated by the designation 10×, 45×, and 96× inscribed on their sides (note that these values may vary somewhat depending upon the particular manufacturer's specifications). The total magnification is obtained by multiplying the magnification of the objective by the magnification of the ocular eyepiece. For example, the total magnification obtained with a 4 mm objective (45×) and a 10× ocular eyepiece is 45 × 10 = 450 diameters.

The highest magnification is obtained with the oil-immersion objective. The bottom tip lens of this objective is very small and admits little light, which is why the iris diaphragm of the condenser must be wide open and the light conserved by means of immersion oil. The oil fills the space between the object and the objective so light is not lost (see figure 2.6 for visual explanation).

Microscope Illumination

Proper illumination is an integral part of microscopy. We cannot expect a first-class microscope to produce the best results when using a second-class illuminator. However, a first-class illuminator improves a second-class microscope almost beyond the imagination. A student microscope with only a mirror (no condenser) for illumination can be operated effectively by employing light from a gooseneck lamp containing a frosted or opalescent bulb. Illuminators consisting of a sheet of ground glass in front of a clear bulb are available but they offer no advantage over a gooseneck lamp. Microscope mirrors are flat on one side and concave on the other. In the absence of a condenser, the concave side of the mirror should be used. Conversely, with a condenser the flat side of the mirror should be used since condensers accept only parallel rays of light and focus them on the slide.

Condensers with two or more lenses are necessary for obtaining the desired numerical aperture. The Abbe condenser, which has a numerical aperture of 1.25, is most frequently used. The amount of light entering the objective is regulated by opening and closing the iris diaphragm located between the condenser and the light source (see figure 2.3). When the oil-immersion objective is used, the iris diaphragm is opened farther than when the high dry or low power objectives are used. Focusing the light is controlled by raising or lowering the condenser by means of a condenser knob.

The mirror, condenser, and objective and ocular lenses must be kept clean to obtain optimal viewing. The ocular lenses are highly susceptible to etching from acids present in body sweat and should be cleaned after each use. (See step 6 below.)

Precautions for Proper Use and Care of the Microscope

Your microscope is a precision instrument with delicate moving parts and lenses. Instruction for proper use and care is as follows:

1. Use both hands to transport the microscope. Keep upright. If inverted, oculars may fall out.
2. Do not touch lenses with your hands. Use lens paper instead. Use of other cleaning materials such as handkerchiefs and Kleenex tissues is discouraged because they may scratch the lens.
3. Do not force the microscope adjustment knobs. If you experience problems making adjustments, consult your instructor.
4. Do not remove objective or ocular lenses for cleaning, or exchange them with different microscopes.
5. For routine cleaning of the oil-immersion objective lens, it is necessary only to wipe off excess oil with a piece of dry lens paper. Any special cleaning should be done under the guidance of the instructor.
6. Before storing the microscope, make certain that the ocular lens is also clean. Frequently, sweat deposits from your eyes, which are acidic, can etch the glass. The presence of other foreign particles can be determined by rotating the ocular lens manually as you look through the microscope. The presence of a

pattern that rotates is evidence of dirt. Clean the upper and lower surfaces of the ocular with lens paper moistened with a drop of distilled water. If dirt persists, consult your instructor. Any dirt remaining after cleaning with a suitable solvent indicates either a scratched lens surface or the presence of dirt on the inside surface of the lens.

7. A blast of air from an air syringe may be effective in removing any remaining dust particles from the lenses.

Definitions

Achromatic objective.   A microscope objective lens in which the light emerging from the lens forms images practically free from prismatic colors.

Apochromatic objective.   A microscope objective lens in which the light emerging from the lens forms images practically free from both spherical and chromatic aberrations.

Bright-field microscopy.   A form of microscopy in which the field is bright and the specimen appears opaque.

Chromatic lens aberration.   A distortion in the lens caused by the different refrangibilities of the colors in the visible spectrum.

Compound microscope. A microscope with more than one lens.

Condenser.   A structure located below the microscope stage that contains a lens and iris diaphragm. It can be raised or lowered, and is used for concentrating and focusing light from the illumination source on the specimen.

Eye relief.   The distance between the outer top lens ocular surface and the point of focus of Ramsden's disk.

Focal length.   The distance from the principal point of a lens to the principal point of focus of the specimen.

Iris diaphragm.   An adjustable opening that can be used to regulate the aperture of a lens.

Magnification.   The ability of a microscope to increase specimen size.

Numerical aperture. A quantity that indicates the resolving power of an objective. It is numerically equal to the product of the index of refraction of the medium in front of the objective lens (n) and the sine of the angle that the most oblique light ray entering the objective lens makes with the optical axis.

Parfocal. Having a set of objectives so mounted on the microscope that they can be interchanged without having to appreciably vary the focus.

Ramsden's disk. A small, real image produced by the ocular lens that can be seen (see text).

Refractive index. The ratio of the velocity of light in the first of two media to its velocity in the second medium as it passes from one medium into another medium with a different index of refraction.

Resolution. The ability of an optical system to form distinguishable images (resolve) of two objects, which are separated by small angular distances.

Simple microscope. A microscope with only one lens.

Spherical lens aberration. An aberration caused by the spherical form of a lens that gives different focal lengths for central and marginal light rays.

Wet mount. A microscope slide preparation in which the specimen is immersed in a drop of liquid and covered with a coverslip.

Working distance. The distance between the tip of the objective lens and the specimen.

Objectives

1. Introduction of historical information on microscopy development from van Leeuwenhoek's single lens light microscope to the compound light microscope of today.
2. Introduction of some major principles of light microscopy, including proper use and care of the microscope.
3. To allow you to use the microscope and become comfortable with it.

References

Dobell, C. *Anton van Leeuwenhoek and his "little animals."* New York: Dover Publications, Inc., 1960.

Gerhardt, P.; Murray, R. G. E.; Costillo, R. N.; Nester, E. W.; Wood, W. A.; Krieg, N. R.; and Phillips, G. B., eds. *Manual of methods for general bacteriology.* Washington, D.C.: American Society for Microbiology, 1981. Contains three excellent chapters on principles of light microscopy.

Gray, P., ed. *Encyclopedia of microscopy and microtechnique.* New York: Van Nostrand-Reinhold, 1973.

Lechevalier, Hubert A., and Solotorovsky, Morris. *Three centuries of microbiology.* New York: McGraw-Hill, 1965. Excellent history of microbiology showing how scientists who made these discoveries were often influenced by other developments in their lives.

Nester et al. *Microbiology: A human perspective.* Chapter 3. Other types of light microscopy are also discussed in this chapter.

Materials

Cake of baker's yeast (sufficient for entire class)

Tube containing 10 ml distilled water (one per student)

Capillary pipet (one per student)

Prepared stained slides of various bacterial forms (coccus, rod, spiral), sufficient for entire class

Procedure

1. Place the microscope on a clear space on your desk, and identify the different parts with the aid of figure 2.3.
2. Before using it be sure to read the "Getting Started" section titled "Precautions for Proper Use and Care of the Microscope."
3. Sample preparation **(wet mount).** Prepare a yeast cell suspension by adding to water in a

test tube just enough yeast to cause visible clouding (approximately 1 loopful per 10 ml of water). Remove a small amount of the suspension with a capillary pipet and carefully place a drop on the surface of a clean slide. Cover the drop with a clean coverslip.

4. Place the wet mount in the mechanical slide holder of the microscope stage with the coverslip side up. Center the coverslip with the mechanical stage control over the stage aperture.

5. Practice focusing and adjusting light intensity when using the low and high power objectives. Rotate the low power objective (10× if available) in position. To focus the objective, you must decrease the distance between the objective lens and the slide. This is done by means of the focusing knobs on the side of the microscope (see figure 2.3). Movement of these knobs on some microscopes causes the objective lens to move up and down in relation to the stage; in other microscopes, the stage moves up and down in relation to the objective. For initial, so-called coarse focusing, the larger adjustment knob is used. For final, so-called fine focusing, the smaller adjustment knob is used. With the large knob, bring the yeast cells into coarse focus. Then complete the focusing process with the fine adjustment knob. Remember that the objective lens should never touch the surface of the slide or coverslip. This precaution is necessary to prevent possible cracking of the slide and scratching the objective lens.

   Adjust the light intensity to obtain optimal image detail by raising or lowering the condenser and by opening or closing the iris diaphragm. For best results, keep the condenser lens at the highest level possible because at lower levels the resolving power is reduced. After examining and drawing a few yeast cells under low power, proceed to the high dry objective by rotating the nosepiece until it clicks into place. If the microscope is parfocal, the yeast cells will already have been brought into approximate focus with the low power so that only fine focusing will be required. Once again, adjust the iris diaphragm and condenser for optimal lighting. If the microscope is not parfocal, it will be necessary, depending on the

type of microscope, either to lower the body tube or to raise the stage with the coarse adjustment knob until it is about $\frac{1}{16}$ inch from the coverslip surface. Repeat these steps to focus the high power objective. Note the increased size of the yeast cells and the decreased number of cells present per microscopic field. Draw a few cells (see Lab Report).

6. Focusing with the oil-immersion objective. First rotate the high dry objective to one side so that a drop of lens immersion oil may be placed on the surface of the coverslip. If the microscope is parfocal, slowly rotate the oil-immersion objective into place. The objective should be in the oil but it should not contact the slide. Complete focusing with the fine adjustment knob, and readjust the iris diaphragm and condenser for optimal lighting. If the microscope is not parfocal, it is necessary to lower the objective into the oil using the coarse adjustment knob (this should not be done while looking through the ocular but while looking from the side of the microscope). Next bring the specimen into coarse focus very slowly with the coarse adjustment knob, and then into sharp focus with the fine adjustment knob. The yeast cells will come into view and go out of view quickly because the depth of focus of the oil-immersion objective is very short. Draw a few cells.

7. Examine the prepared stained bacteria slides with the oil-immersion objective. (See exercise 4, Procedure, Simple Stain step 11 for information on how to prepare and focus stained slides with the oil-immersion objective.) Once again, if your microscope is parfocal, first focus the slide with the lower power objective before using the oil-immersion objective. Draw a few cells of each bacterial form. Compare the shapes of these cells with those in color plates 3–5.

8. When you finish this procedure, wipe the excess oil from the oil-immersion objective with lens paper, and if necessary clean the ocular (see "Precautions for Proper Use and Care of the Microscope"). Next return the objective to the low power setting, and if your microscope has an adjustable body tube, lower (rack down) it before returning the microscope to the microscope cabinet.

2–8   *Exercise 2 Bright-field Light Microscopy, Including History and Working Principles*

EXERCISE 2

Laboratory Report: Bright-field Light Microscopy, Including History and Working Principles

Results

1. Draw a few yeast cells from each magnification. Include any interesting structural changes evident at the three magnifications.

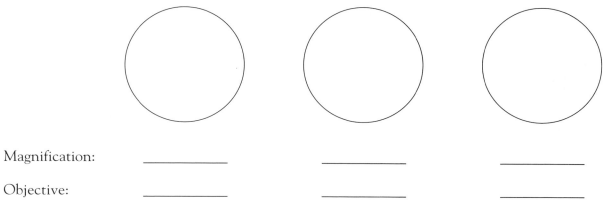

Magnification: _____   _____   _____

Objective: _____   _____   _____

2. Examination of prepared bacteria slides. Examine with the oil-immersion objective and draw a few cells of each morphological form.

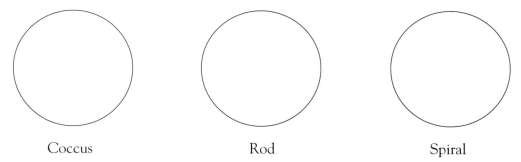

    Coccus   Rod   Spiral

3. Answer the following questions about your microscope:
   a. What is the magnification and numerical aperture (NA) stamped on each objective of your microscope?

   Objective Magnification   **Numerical Aperture**

   _____   _____

   _____   _____

   _____   _____

   _____   _____

b. What is the magnification stamped on the oculars? _____

c. Calculate the total magnification of the objective/ocular combination with:

The lowest power objective: _____

The highest power objective: _____100_____

Questions

1. Discuss the advantages of a modern compound microscope (figure 2.3) over an early compound microscope (figure 2.2).

   _____

   _____

   _____

2. Why must the distance from slide to objective increase rather than decrease when coarse focusing with the high dry and oil-immersion objectives?

   _____

3. How does increasing the magnification affect the resolving power?

   _____

4. How does lens immersion oil help to increase the resolving power of the oil-immersion objective?

   _____

   _____

5. Distinguish the real image from the retinal image. Which image is seen by your eye?

   _____

   _____

6. What are the functions of the substage condenser?

   _____

   _____

7. What is meant by the term "parfocal"? Does it apply to your microscope?

   _____

True-False Questions

Mark the statements below true (T) or false (F).

1. Van Leeuwenhoek's microscope was corrected for spherical but not chromatic aberrations. _____

2. Spherical lens aberrations are easier to correct than chromatic lens aberrations. _____

3. The objective NA is more important than the condenser NA for increasing resolving power. _____

4. The working distance is the distance from the tip of the objective to the tip of the condenser lens. _____

5. The image produced by the ocular lens known as Ramsden's disk is in focus outside of the ocular. _____

NOTES: Hanging drop technique: use depression slide

Candida albicans
Bacillus subtilus – rod
Staphylococcus epidernitis – coccus
Proteus vulgaris – rod

Methylene blue - stains dead cells, living cells break it down and appear colorless.

Star diaphragm - blocks light in dark field microscope
— moving objects pick up light

EXERCISE 3

Microscopic (Bright-field and Dark-field) Determination of Cell Motility, Form, and Viability Using Wet Mount and Hanging Drop Preparations

Getting Started

Although cell motility is usually determined by the semisolid agar stab inoculation method, it is sometimes determined by direct microscopic examination. Microscopic examination allows for the determination of the form of cells, for example, their general shape (round or coccus, elongate or rod, etc.); and their arrangement, for example, how the cells adhere and attach to one another (as filaments, tetrads, etc.). Finally, it is also sometimes possible to determine cell viability using either bright-field microscopy and a vital stain or dark-field microscopy without a stain. With dark-field microscopy, living cells appear bright and dead cells appear dull. With bright-field microscopy and methylene blue, living cells appear colorless, whereas dead cells appear blue. The dead cells are unable to enzymatically reduce methylene blue to the colorless form.

For all of these methods, a **wet mount slide** or a **hanging drop slide** cell preparation is used. Wet mounts are easier to prepare but dry out more rapidly due to contact between the coverslip and air on all four sides. (Drying out can be reduced by ringing the coverslip edges with petroleum jelly.) Other disadvantages are the inability at times to see the microorganism because it is not sufficiently different in refractive index from the suspending fluid (this can sometimes be resolved by reducing the light intensity), and it is not particularly useful for observing thick preparations such as hay infusions.

In this exercise, bright-field microscopy is used with wet mounts to observe bacterial motility and form. In observing bacterial motility, it is important to distinguish true motility from "Brownian movement," a form of movement caused by molecules in the liquid striking a solid object, in this instance the bacterial cell, causing it to vibrate back and forth. If the bacterial cell is truly motile, you will observe its directional movement from point A to point B, providing the cells are not in a **resting stage** of the growth curve.

Measurement of cell viability with methylene blue may also be skewed when resting stage cells are used (Kleyn et al., 1962). Resting stage cells, although viable, are often unable to reduce the dye to a colorless form. Thus, it is preferable to observe cells from the early logarithmic stage of the growth curve (see exercise 10). The cells of choice—yeast—are sufficiently large for ease of observation with bright-field microscopy when using the high dry objective. Unstained cells from the same stage of the growth curve will also be observed for viability by using dark-field microscopy. Thus, you will be able to compare your viability results with one another. Hopefully they will vary no more than 10%—one accepted standard of error for biological material.

Finally, you will have an opportunity to observe bacterial motility by means of a hanging drop slide preparation.

Definitions

Dark-field microscopy. A form of microscopy in which the specimen is brightly illuminated on a dark background.

Depression slide. A microscope slide with a circular depression in its center.

Hanging drop slide. A specimen observation technique in which the specimen hangs suspended from an inverted coverslip mounted on a depression slide.

Resting stage. The stage of the growth curve in which cells are metabolically inactive.

Star diaphragm. A metal diaphragm used for dark-field microscopy. Its opaque center deflects the light rays that converge on the objective so that only the oblique rays strike the specimen. The net result is a dark-colored microscope field with a brightly colored specimen.

Figure 3.1

Conversion of a bright-field microscope into a dark-field microscope by inverting a star diaphragm into the filter holder located below the condenser lens. Courtesy of Dr. Harold J. Benson

Vital stain.   A stain able to differentiate living from dead cells, e.g., methylene blue is colorless when reduced in the presence of hydrogen, while remaining blue in its absence.

Wet mount slide.   A microscopic specimen observation technique in which a drop containing the specimen is placed on the surface of a clean slide, followed by the addition of a coverslip over the drop.

Objectives

1. To become familiar with the advantages and limitations of wet mount and hanging drop preparations for observing living cell material. This will be achieved both by reading and direct experience using living bacteria and yeast cultures as specimen material.
2. To learn how to use dark-field microscopy to observe living cells.

References

Kleyn, J.; Mildner, R.; and Riggs, W. 1962. Yeast viability as determined by methylene blue staining. *Brewers Digest* 37 (6):42–46.

Nester et al. *Microbiology: A human perspective*. Chapters 1 and 3.

Materials

Cultures

12–18 hour nutrient broth cultures of *Staphylococcus epidermidis*, and *Spirillum volutans* showing visible clouding

12–18 hour nutrient broth cultures of *Bacillus cereus* and *Pseudomonas aeruginosa* showing visible clouding

A yeast suspension previously prepared by suspending sufficient baker's yeast in a tube of glucose yeast fermentation broth to produce visible clouding, followed by 6–8 hour incubation at 25° C

A hanging drop depression slide

Vaseline and toothpicks

Pasteur pipets

Dropper bottle with acidified methylene blue

A star diaphragm for dark-field microscopy (figure 3.1)

Procedure

Wet Mounts for Study of Bacterial Form and Motility

1. Prepare six clean microscope slides and seven clean coverslips by washing them in a mild detergent solution, rinsing with distilled water, and then drying them with a clean towel. Examine visually for clarity.
2. Suspend your broth culture of *S. epidermidis* by gentle tapping on the outside of the culture

tube. Hold the tube firmly between thumb and index finger and tap near the bottom of the test tube with your finger until the contents mix.

3. Remove the test-tube cover and with a Pasteur pipet, finger pipette approx. 0.1 ml of the broth culture.

4. Transfer a drop of this suspension to the surface of a slide.
   Note: The drop must be of suitable size; if it is too small, it will not fill the space between the coverslip and the slide; if it is too large, some of the drop will pass outside the coverslip, which could smear the front lens of the microscope objective.
   Discard the Pasteur pipet in the designated container.

5. Grasp a clean coverslip on two edges and place it carefully over the surface of the droplet.

6. Insert the wet mount on the stage of your microscope and examine for cell motility and form with the oil-immersion objective. Make certain you can distinguish true motility from Brownian movement. Prepare a drawing of some of the cells and record your findings in the Laboratory Report.

7. Discard the slide in the designated container for autoclaving.

8. Repeat the above procedure with *S. volutans*, *B. cereus*, and *P. aeruginosa* (for representative cell shapes see color plates 3–5).

Use of Hanging Drop Slides for Study of Bacterial Form and Motility

1. Prepare a clean depression drop slide and coverslip.

2. With a toothpick, spread a thin ring of Vaseline approximately ¼ inch outside the depression slide concavity (figure 3.2*a*).

3. Using your suspended *B. cereus* broth culture and a wire loop, transfer 2 loopfuls to the central surface of a coverslip (see figure 3.2*b*).

4. Invert the depression slide and center the depression over the droplet on the coverslip. Make contact and press lightly, forming a seal between the Vaseline ring and coverslip (see figure 3.2*c*).

Figure 3.2 (*a–c*) Preparation of a hanging drop slide.

(a) Depression slide

(b) Coverslip

(c) Pressing of slide against cover glass

5. Quickly turn over the depression slide so as not to disrupt the culture droplet.
   Note: If done correctly, the droplet will remain suspended and will not come in contact with the well bottom.

6. Place the slide on the stage of your microscope and first focus the *edge* of the droplet with your low power objective. You may also need to reduce the light to achieve proper contrast. Due to capillary action, most microorganisms gather at the edge. When in focus, the edge will appear as a light line against a dark background.

7. In order to see individual bacterial cells, you will need to use the oil-immersion objective. Add a drop of lens immersion oil to the coverslip, and if parfocal, shift to the oil-immersion objective. Once again, light adjustment becomes necessary. You should now be able to observe individual bacteria, their form, and motility. If not, ask your instructor for help.

8. Draw some of the cells and record their motility and other findings in the Laboratory Report. Discard slide in waste glass container.

Use of Dark-field Microscopy to Determine Yeast Cell Viability

1. Insert the star diaphragm into the filter holder located below the microscope condenser (see figure 3.1).
   Note: Make certain that it is accurately centered.

2. Suspend the baker's yeast preparation and prepare a wet mount. Transfer the wet mount to the microscope stage.

3. Examine the wet mount with the low power objective. Keep the iris diaphragm wide open in order to admit as much light as possible.

4. Adjust the condenser focus to the position where the best dark-field effect is obtained. See color plate 6 for examples of yeast photographed with bright-field and dark-field microscopy.

5. Examine the wet mount with the high dry objective.
   Note: Dark-field microscopy may or may not be possible at this magnification depending upon how well the oblique light rays pass through the objective lens.

6. Determine the percent of viable yeast cells. To do so, count a total number of approximately 100 or more cells and also the number of dull-looking cells (dead cells) within this total. With this information, you can calculate the percent of viable yeast cells.

7. In the Laboratory Report, prepare drawings of representative cells and show your calculations for determining the percent of viable cells.

Use of a Vital Stain, Methylene Blue, to Determine Yeast Viability

1. From a dropping bottle, transfer a small drop of methylene blue to the surface of a clean slide.

2. With a Pasteur pipet, add a small drop of the baker's yeast suspension. Carefully place a clean coverslip over the surface of the droplet.

3. Observe the wet mount with bright-field microscopy using the low and high dry microscope objectives.

4. For the Laboratory Report, prepare drawings of representative cells and show your calculations for determining the percent of viable yeast cells. In this instance, dead cells will be stained blue and viable cells will remain colorless.

EXERCISE

3

Laboratory Report: Microscopic (Bright-field and Dark-field) Determination of Cell Motility, Form, and Viability Using Wet Mount and Hanging Drop Preparations

Results

1. Wet Mounts for Study of Bacterial Form and Motility
   Drawings of representative cells showing their relative sizes, shapes, and arrangements. Record magnification and motility (+ or −).

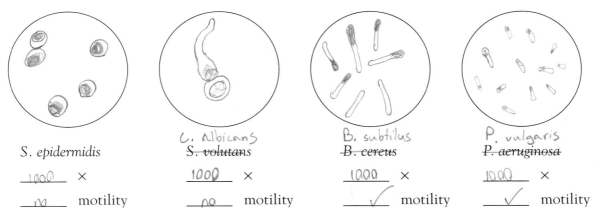

S. epidermidis	*C. Albicans* ~~*S. volutans*~~	*B. subtilus* ~~*B. cereus*~~	*P. vulgaris* ~~*P. aeruginosa*~~
_1000_ ×	_1000_ ×	_1000_ ×	_1000_ ×
_no_ motility	_no_ motility	_✓_ motility	_✓_ motility

2. Hanging Drop Slide (*B. cereus*)
   Make observations similar to those above and indicate any differences from the *B. cereus* wet mount observations.

   Differences: _____

   _____

   _____

B. cereus
_400_ ×
_✓_ motility

3. Dark-field Microscopy of Baker's Yeast
   Drawings of cells showing their size, shape, and arrangement, as well as the visual appearance of living and dead cells. Record magnification used.

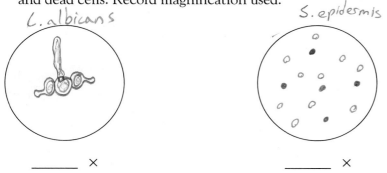

_____ ×   _____ ×

   Show your calculations for determining the percent of viable cells.

4. Bright-field Microscopy of Baker's Yeast Stained with Methylene Blue
   Make the same kind of observations as in number 3.

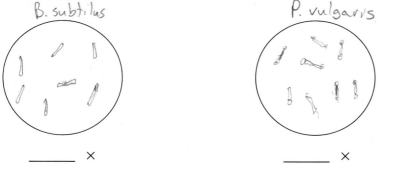

_____ ×   _____ ×

   Show your calculations for determining the percent of viable cells.

   Discuss your yeast cell viability results by the above two methods. If a wide viability variance (>10%) exists between the two methods, what other method might you use to prove which method is more accurate? You may wish to consult your text (chapter 3) or lab manual (exercise 9) for help in constructing a reasonable answer.

Questions

1. What advantages are there in determining cell motility microscopically rather than with a stab culture?

2. What advantages does a hanging drop preparation have over a wet mount preparation? Disadvantages?

3. How did you obtain optimal results with dark-field microscopy?

4. Why is it difficult to employ the oil-immersion objective for dark-field microscopy?

5. What might be one reason for employing an actively multiplying culture when examining viability microscopically?

6. In addition to determining cell viability, what other useful morphological determination can sometimes be made with dark-field microscopy? Consult your text.

7. What difficulties might there be in attempting to determine the viability of bacterial cells with stains such as methylene blue? This will no doubt require some investigation of the literature. A possible clue lies in the prokaryotic makeup of bacteria. Yeasts, on the other hand, are eukaryotic cells.

I NTRODUCTION to Staining of Microorganisms

Bacteria are difficult to observe in a broth or wet mount because there is very little contrast between them and the liquid in which they are suspended. This problem is solved by staining bacteria with dyes. Although staining kills bacteria so their motility cannot be observed, the stained organisms contrast with the surrounding background and are much easier to see. The determination of the shape, size, and arrangement of the cells after dividing are all useful in the initial steps in identifying an organism. These can be demonstrated best by making a smear on a glass slide from the clinical material, broth culture, or colony from a plate, then staining the smear with a suitable dye. Examining a stained preparation is one of the first steps in identifying an organism.

Staining procedures used here can be classified into two types: the simple stain and the multiple stain. In the simple stain, a single stain such as methylene blue or crystal violet is used to dye the bacteria. The shape and the grouping of the organisms can be determined, but all organisms (for the most part) are stained the same color. Another kind of simple stain is the negative stain. In this procedure, the organisms are mixed with a dye and permitted to dry. When they are observed, the organisms are clear against a dark background.

The multiple stain involves more than one stain. The best known example is the Gram stain, which is widely used. After staining, some organisms appear purple and others pink, depending on the structure of their cell wall.

Multiple stains are frequently known as differential stains because they are used to visualize special structures of bacteria. In contrast with eukaryotic organisms, prokaryotic organisms have relatively few morphological differences. Several of these structures such as endospores, capsules, acid-fast cell walls, storage bodies, and flagella can be seen with special stains. In the next two exercises, you will have an opportunity to stain bacteria with a variety of staining procedures and observe these structures.

To a person uninstructed in natural history, his country or seaside stroll is a walk through a gallery filled with wonderful works of art, nine-tenths of which have their faces turned to the wall.

Thomas Henry Huxley (1873)

methylene blue - stains dead bacteria

crystal violet -

Safrinin

Malachite Green

Bacillus subtilus

Staphylacaccus epidermidis

EXERCISE 4

Simple Stains: Positive and Negative Stains

Getting Started

Two kinds of single stains will be done in this exercise: the simple stain and the negative stain. Microbiologists most frequently stain organisms with the Gram stain, but in this exercise a simple stain will be used to give you practice staining and observing bacteria before doing the more complicated multiple, or differential, stains.

After you have stained your bacterial smears, you can examine them with the oil-immersion lens, which will allow you to distinguish the morphology of different organisms. The typical bacteria you will see are about 0.5–1.0 micrometer (μm) in width to about 2–7 μm long and are usually rods, cocci, or spiral-shaped. Sometimes rods are referred to as bacilli, but since that term is also a genus name (*Bacillus*) for a particular organism, the term rod is preferred.

Another kind of simple stain is the negative stain. Although it is not used very often, it is advantageous in some situations. Organisms are mixed in a drop of nigrosin or India ink on a glass slide. After drying, the organisms can then be observed under the microscope as clear areas in a black background. This technique is sometimes used to observe capsules or inclusion bodies. It also prevents eyestrain when many fields must be scanned. The dye tends to shrink away from the organisms, causing cells to appear larger than they really are.

In both of these simple stains, you will be able to determine the shape of the bacteria and the characteristic grouping after cell division (as you did in the wet mounts). Some organisms tend to stick together after dividing and form chains or irregular clumps. Others are usually observed as individual cells. However, this particular characteristic depends somewhat on how the organisms are grown. *Streptococcus* form long, fragile chains in broth, but if they grow in a colony on a plate, it is sometimes difficult to make a smear with these chains intact.

Definitions

Differential stain. A procedure that stains specific morphological structures—usually a multiple stain.

Inclusion bodies. Granules of storage material such as sulfur that accumulate within some bacterial cells.

Micrometer. (abbreviated μm) The metric unit used to measure bacteria. It is 10^{-6} m (meter) and 10^{-3} mm (millimeter).

Negative stain. A simple stain in which the organisms appear clear against a dark background.

Parfocal. If one objective lens of a microscope is in focus, all lenses will be in focus when used.

Simple stain. A procedure for staining bacteria consisting of a single stain.

Smear. A dried mixture of bacteria and water (or broth) on a glass slide in preparation for staining.

Objectives

1. Learn to prepare and stain a bacterial smear using a simple stain.
2. Observe stained organisms under the oil-immersion lens.
3. Prepare and observe a negative stain.
4. Observe the various morphologies and arrangements of bacteria in stained preparations.

References

Gerhardt, Philip, ed. *Manual for general and molecular bacteriology.* American Society for Microbiology, Washington, D.C., 1994.

Nester et al. *Microbiology: A human perspective.* Chapter 3, 51–54.

Materials

Cultures
   Bacillus subtilis or *B. cereus*
   Staphylococcus epidermidis
   Enterococcus faecalis
   Escherichia coli

Staining bottles with:
   crystal violet
   methylene blue
   safranin

Glass slides

Wax pencils or waterproof marking pen

Tap water in small dropper bottle (optional)

Inoculating loop

Alcohol sand bottle (a small screw cap bottle half full of sand and about three-quarters full of 95% alcohol)

Alcohol bottle and inoculating loop

Procedure

Simple Stain

1. Clean a glass slide by rubbing it with slightly moistened cleansing powder such as Boraxo or Bon Ami. Rinse well and dry with a paper towel. Even new slides should be washed because sometimes they are covered with a protective coating.

2. Draw two or three circles with a waterproof pen or wax pencil on the underside of the slide. If the slide has a frosted portion, you can also write on it with a pencil. This is useful because it is easy to forget the order in which

Figure 4.1 Slide with three drops of water. Three different bacteria can be stained on one slide.

you placed the organisms on the slide and you can list them, for instance, from left to right (figure 4.1).

3. Add a drop of water to the slide on top of each of the circles. Use your loop to transfer tap water or use water from a dropper bottle. This water does not need to be sterile. Although there are some organisms (nonpathogens) in municipal water systems, there are too few to be seen.

    If you are preparing a smear from a broth culture as you will do in the future, add only the broth to the slide. Broth cultures are relatively dilute, so no additional water is added.

4. Sterilize a loop by holding it at an angle in the flame of the Bunsen burner. Heat the entire wire red hot, but avoid putting your hand directly over the flame or heating the handle itself (figure 4.2).

5. Hold the loop a few seconds to cool it, then remove a small amount of a bacterial culture and suspend it in one of the drops of water on the slide (see figure 4.2). Continue to mix in bacteria until the drop becomes slightly turbid (cloudy). If your preparation is too thick, it will stain unevenly and if it is too thin you will have a difficult time finding organisms under the microscope. In the beginning, it may be better to err on the side of having a slightly too turbid preparation— at least you will be able to see organisms and you will learn from experience how dense to make the suspension.

6. Heat the loop red hot. It is important to burn off the remaining organisms so that you will not contaminate your bench top. If you rest your loop on the side of your Bunsen burner, it can cool without burning anything on the bench.

    Sometimes the cell material remaining on the loop spatters when heated. To prevent this, some laboratories remove bacterial cell

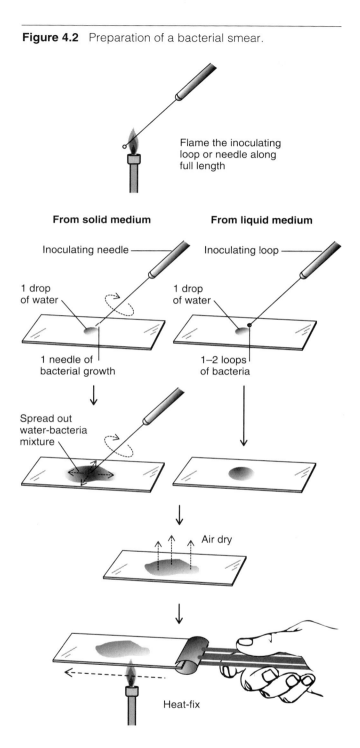

Figure 4.2 Preparation of a bacterial smear.

Flame the inoculating loop or needle along full length

From solid medium

Inoculating needle

1 drop of water

1 needle of bacterial growth

Spread out water-bacteria mixture

From liquid medium

Inoculating loop

1 drop of water

1–2 loops of bacteria

Air dry

Heat-fix

material from the loop by dipping the loop in a bottle of sand covered with alcohol. Then the loop is heated red hot in the Bunsen burner.

7. Permit the slide to dry. Do not heat it in any way to hasten the process, since the cells will become distorted. Place the slide off to the side of the bench so that you can proceed with other work.

8. When the slide is dry (in about 5–10 minutes), heat-fix the organisms to the slide by quickly passing it through a Bunsen burner flame two or three times so that the bottom of the slide is barely warm. This step causes the cells to adhere to the glass so they will not wash off in the staining process (figure 4.3).

9. Place the slide on a staining bar over a sink or pan. Alternatively, hold the slide over the sink with a forceps or clothespin. Cover the specimen with a stain of your choice—crystal violet is probably the easiest to see (figure 4.3).

10. After about 20 seconds, pour off the stain and rinse with tap water (figure 4.3).

11. Carefully blot the smear dry with a paper towel. Do not rub the slide from side to side as that will remove the organisms. Be sure the slide is completely dry (figure 4.3).

12. Observe the slide under the microscope. Since you are looking at bacteria, you must use the oil-immersion lens in order to see them. One method is to focus the slide on low power, then cover the smear with immersion oil and move the immersion lens into place. If your microscope is parfocal, it should be very close to being in focus. Note that no coverslip is used when looking at stained organisms.

Another method for focusing the oil-immersion lens is to put oil on the smear, and then while looking at the microscope from the side very carefully raise the stage (or lower the lens, depending on your microscope) until the immersion lens is just barely touching the slide. Then when looking through the microscope, very slowly back the lens off the slide until it is

Figure 4.3 (*a*) Staining, (*b*) washing, and (*c*) blotting a simple stain. From John P. Harley and Lansing M. Prescott, *Laboratory Exercises in Microbiology,* 2d ed. Copyright © 1993 Wm. C. Brown Communications, Inc., Dubuque, Iowa. All Rights Reserved. Reprinted by permission.

Simple staining procedure

Staining loop

Sink or suitable receptacle

Staining bottle

(a)

Water

Wash bottle

(b)

Gentle blotting

(c)

in focus. Never move the immersion lens toward the slide while looking through the microscope. You may hit the slide with the lens and damage the lens. When you have a particularly thin smear, it is sometimes helpful to put a mark on the slide near the stain with a marking pen. It is easy to focus on the pen mark, and you will know that you have the top of the slide in focus and can then search for the smear.

13. Record your results.
14. If you want to save your stained slide, it can be saved with the oil on it. If you do not want to save the slide, simply clean it with cleanser and water. The staining procedure kills the bacteria and the slide does not need to be boiled before cleaning.
15. Wipe off the oil from the immersion lens with lens paper before storing the microscope.

The Negative Stain

This stain can be used to observe capsules or storage material. However, in this exercise the negative stain will be used to compare the appearance of the same organisms using the two staining procedures.

Materials

Culture
Same cultures used for simple stain
Bottle of India ink

Procedure

1. Place a drop of water on a clean slide and add organisms with a loop until the drop is cloudy.
2. Mix a loopful of India ink into the drop and spread the mixture out into a thin film.
3. Let dry and examine under the microscope. Bacteria can be seen as clear areas on a black background.
4. Record your results.

EXERCISE

4

Laboratory Report: Simple Stains: Positive and Negative Stains

Results

1. Simple stain

	Staphylococcus	*Bacillus*	*Escherichia*	*Enterococcus*
Shape and arrangement	red safranin crystal violet (purple)	red safranin		

2. Negative stain — malachite green (green)

methylene blue (blue)

How did the organisms appear?

Questions

1. What are the advantages of a simple stain over a wet mount?

2. Do you need more or less light when viewing a stained preparation compared to a wet mount?

3. What information can you observe in a wet mount that cannot be seen in a stained preparation?

4. How does the negative stain compare to the simple stain?

[handwritten notes]

1 Finger
2 shoe
3 saliva swab

2 TSA plates
1,10,13,9
37°C → Room Temp

entire ◯

serrate ᗞᗞᗞᗞ

✳ filamentous

entire *(irregular shape)*

undulate

lobate ᔕᔕᔕ

micrococcus luteus
↳ yellow (C)

Times are important
crystal violet – 1 min → rinse
iodine – 1 minute → rinse
acetone alcohol – 5 sec → rinse
safranin – 1 minute → rinse

EXERCISE 5

Multiple and Differential Stains

Getting Started

Multiple stains involve at least two dyes. They are also called differential stains because they specifically stain certain morphological features.

Gram Stain

The Gram stain is especially useful as one of the first procedures in identifying organisms because it reveals not only the morphology and the arrangement of the cells, but also information about the cell wall.

Near the turn of the century, Christian Gram devised the staining procedure when trying to stain bacteria so that they contrasted with the tissue sections he was observing. Many years later, it was found that purple (gram-positive) bacteria had thick cell walls of peptidoglycan, while pink (gram-negative) bacteria had much thinner cell walls of peptidoglycan surrounded by an additional membrane. The thick cell wall retains the purple dye in the procedure, but the thin wall does not.

In the Gram stain, a bacterial smear is dried and then heat-fixed to cause it to adhere to the glass slide (as in the simple stain). It is then stained with crystal violet dye, which is rinsed off and replaced with an iodine solution. The iodine acts as a mordant—that is, it binds the dye to the cell. The smear is then decolorized with alcohol and counterstained with safranin. In gram-positive organisms, the purple crystal violet dye, complexed with the iodine solution, is not removed by the alcohol and thus the organisms remain purple. On the other hand, the purple stain is removed from gram-negative organisms by the alcohol and the colorless cells take up the red color of the safranin counterstain.

Note: Many clinical laboratories use a 50/50 mixture of alcohol and acetone because it destains faster than 95% alcohol. If the instructor would rather not use acetone, 95% alcohol is just as effective, but the stain must be decolorized longer (up to 30 seconds).

Appearance of the Cells after Each Procedure		
	Gram +	Gram–
Crystal violet *primary*	Purple	Purple
Iodine (*mordant*)	Purple	Purple
Alcohol *acetone*	Purple	Colorless
Safranin (*counter*)	Purple	Pink

→ dissolves LPS

Special Notes to Improve Your Gram Stains

1. Gram-positive organisms can lose their ability to retain the crystal violet complex when they are old. This can happen when a culture has only been incubating 18 hours—the genus *Bacillus* is especially apt to become gram negative. Use young, overnight cultures whenever possible. It is interesting to note that gram-positive organisms can appear gram negative, but gram-negative organisms almost never appear gram positive.

2. Another way gram-positive organisms may appear falsely gram negative is by over decolorizing in the Gram-stain procedure. If excessive amounts of acetone/alcohol are used, almost any gram-positive organism will lose the crystal violet stain and appear gram negative.

3. If you are staining a very thick smear, it may be difficult for the dyes to penetrate properly. This is not a problem with broth cultures, which are naturally quite dilute, but be careful not to make the suspension from a colony in a drop of water too thick.

4. When possible, avoid making smears from inhibitory media such as Eosin Methylene Blue (EMB) because the bacteria frequently give variable staining results and can show atypical morphology.

5. The use of safranin in the Gram stain is not essential. It is simply used as a way of dying the colorless cells so they contrast with the purple. For those who are color-blind and have difficulty distinguishing pink from purple, other dyes might be tried as counterstains.

Definitions

Counterstain. A stain used to dye unstained cells a contrasting color in a differential stain.

Mordant. A substance that increases the adherence of a dye.

Peptidoglycan. The macromolecule making up the cell wall of most bacteria.

Vegetative cell. A cell that has not formed spores or other resting stages.

Objectives

1. To learn the Gram-stain procedure.
2. To learn to distinguish gram-positive organisms from gram-negative organisms.

References

Gerhardt, Philip, ed. *Manual for general and molecular bacteriology.* American Society for Microbiology, Washington, D.C., 1994.

McGonagle, Lee Anne. *Procedures for diagnostic bacteriology.* 7th ed. Seattle, WA. Department of Laboratory Medicine, University of Washington, 1992.

Nester et al. *Microbiology: A human perspective.* Chapter 3, 52–54.

Materials

Staining bottles of the following:
    crystal violet
    iodine
    acetone/alcohol or 95% alcohol
    safranin

Clothespin or forceps

Staining bars

Overnight cultures growing on TS agar slants
    Escherichia coli
    Bacillus subtilis
    Staphylococcus epidermidis
    Enterococcus faecalis
    Micrococcus luteus

Procedure

1. Put two drops of water on a clean slide. In the first drop, make a suspension of the unknown organism to be stained just as you did for a simple stain (see preparing a smear in exercise 4). In the second drop, mix together a known gram-positive organism and a known gram-negative organism. This mixture is a control to ensure that your Gram-stain procedure (figure 5.1) is giving the proper results. Heat fix the slide.

2. Place the slide on a staining bar across a sink (or can). Alternatively, hold the slide with a clothespin or forceps over a sink.

3. Flood the slide with crystal violet until the slide is completely covered. Leave it on for 6–30 seconds and then discard into the sink. The timing is not critical. Rinse the slide with water from a wash bottle or gently running tap water.

4. Flood the slide with Gram's iodine for about 12–60 seconds and wash with water.

5. Hold the slide at a 45° angle and carefully drip acetone ethanol over it until no more purple dye runs off. Immediately wash slide with tap water. Thicker smears may take longer than thinner ones, but acetone/alcohol should usually be added for 1–2 seconds and no more than 5 seconds. Timing is critical in this step.

6. Flood the slide with safranin and leave it on for 10–30 seconds—timing is not important. Wash with tap water. Safranin is a counterstain because it stains the cells that have lost the purple dye.

7. Blot the slide carefully with a paper towel to remove the water, but do not rub from side to side. When it is completely dry, observe the slide under the microscope. Remember that you must use the oil-immersion lens to observe bacteria. Compare your stain to the control mixture on the same slide and with color plate 7.

8. Describe the appearance of your stained bacteria in the "Results" section.

Figure 5.1 (a–i) Gram-stain procedure. From John P. Harley and Lansing M. Prescott, *Laboratory Exercises in Microbiology,* 2d ed. Copyright © 1993 Wm. C. Brown Communications, Inc., Dubuque, Iowa. All Rights Reserved. Reprinted by permission.

(a) Crystal violet; 6–30 seconds

(b) Rinse for 5 seconds

(c) Cover with Gram's iodine; 12–60 seconds

(d) Rinse with water for 5 seconds

(e) Decolorize for 1–5 seconds

(f) Rinse with water for 5 seconds

(g) Counterstain with safranin; 10–30 seconds

(h) Rinse for 5 seconds

(i) Blot dry with a paper towel

9. Be sure to remove the immersion oil from the lens with lens paper before storing the microscope.

Getting Started

Optional Stains

Acid-fast Stain

The acid-fast stain is useful for identifying bacteria with a waxy lipid cell wall. Most of these organisms are members of a group of bacteria called mycobacteria. Although there are many harmless bacteria in this group, it also includes *Mycobacterium tuberculosis*, which is the cause of tuberculosis in humans. These organisms have a gram-positive cell wall structure, but the lipid in the cell wall prevents staining with the Gram-stain dyes.

In the Ziehl-Neelsen acid-fast stain procedure, the carbolfuchsin dye is driven into the waxy cell wall with steam. Once the lipid-covered cell has been dyed, it cannot easily be decolorized—even with alcohol containing HCL (which is called acid alcohol). Nonmycobacteria are also dyed with the carbolfuchsin, but are decolorized by acid alcohol. These colorless organisms are stained with methylene blue so they contrast with the pink acid-fast bacteria that were not decolorized.

The reason this stain is important is that one of the initial ways tuberculosis can be diagnosed is by the presence of *Mycobacterium* in a patient's sputum. (Sputum is a substance that is coughed up from the lungs and contains puslike material.) Tuberculosis is a very serious disease worldwide and is starting to increase in the United States after decreasing for about 80 years. Since the process of finding acid-fast organisms in sputum is quite difficult and time-consuming, this test is usually performed in state health laboratories.

Objectives

1. To become familiar with acid-fast organisms.
2. To prepare an acid-fast stain.

References

Gerhardt, Philip, ed. *Manual for general and molecular bacteriology.* American Society for Microbiology, Washington, D.C., 1994.

Nester et al. *Microbiology: A human perspective.* Chapter 3, 54.

Materials

Cultures
  Mycobacterium smegmatis
Carbolfuchsin in staining bottles
Methylene blue in staining bottles
Acid alcohol in staining bottles
Beaker
Metal or glass staining bars

Procedure

1. Prepare a smear of the material and heat-fix (see exercise 4).
2. Place the slide on staining bars over a beaker of boiling water, and cover the smear with a piece of paper towel torn to fit the slide (figure 5.2). Cover the slide with carbolfuchsin. The paper will help keep the dye on the slide as you are steaming it.
3. Keep the slide covered with stain and steam for at least 5 minutes.
4. Remove paper and wash off the carbolfuchsin in tap water.
5. Flush all freely removable stain with acid alcohol (not ordinary alcohol).
6. Flood the entire slide with acid alcohol and allow to stand for a minute or two.
7. Wash with water for 5 seconds.
8. Counterstain for about 30–45 seconds with methylene blue.
9. Wash with water.
10. Blot dry carefully and examine under the oil-immersion lens.
11. Record results.

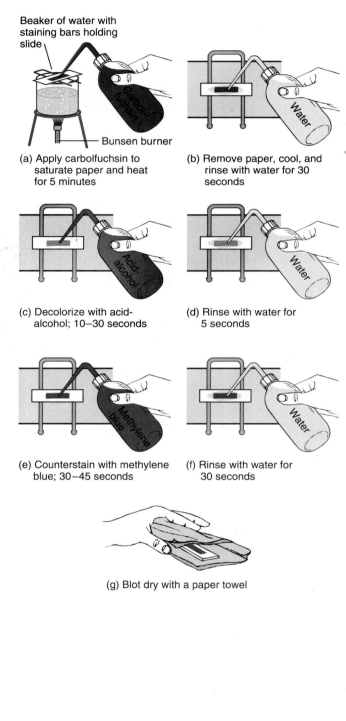

Figure 5.2 (*a–g*) Acid-fast staining procedure. From John P. Harley and Lansing M. Prescott, *Laboratory Exercises in Microbiology,* 2d ed. Copyright © 1993 Wm. C. Brown Communications, Inc., Dubuque, Iowa. All Rights Reserved. Reprinted by permission.

(a) Apply carbolfuchsin to saturate paper and heat for 5 minutes

(b) Remove paper, cool, and rinse with water for 30 seconds

(c) Decolorize with acid-alcohol; 10–30 seconds

(d) Rinse with water for 5 seconds

(e) Counterstain with methylene blue; 30–45 seconds

(f) Rinse with water for 30 seconds

(g) Blot dry with a paper towel

Getting Started

Differential Stains of Bacterial Cell Structures

Although bacteria have few cell structures observable by light microscopy when compared to other organisms, some have:

1. Capsules. A capsule is a somewhat gelatinous coating surrounding the cell. It can consist of amino acids or carbohydrates and it can protect the bacterium from engulfment by white blood cells. The ability to produce a capsule frequently depends on the availability of certain sugars. *Streptococcus mutans,* for example, produces a capsule when growing on sucrose, but not when growing on glucose.

2. Endospores. Some organisms such as *Bacillus* and *Clostridium* can form a resting stage called an endospore, which will protect them from heat, chemicals, and starvation. When the cell determines that conditions are becoming unfavorable due to a lack of nutrients or moisture, it forms an endospore. Then when conditions become favorable again the spore can germinate and the cell can continue to divide. The endospore is resistant to most stains so special staining procedures are needed.

3. Storage granules. Some organisms have storage granules of phosphate, sulfur, or carbohydrate. Some of these granules can easily be seen with certain stains.

4. Bacterial flagella. Some bacteria have flagella (flagellum, singular) for motility. Their width is below the resolving power of the microscope so they cannot be seen in a light microscope (the flagella seen at each end of *Spirillum* in a wet mount is actually a tuft of flagella). Flagella can be visualized if they are dyed with a special stain that precipitates on them, making them appear much thicker. The arrangement of the flagella on bacteria is usually characteristic of the organism and can aid in identification.

Objectives

1. To become familiar with various structures and storage products of bacteria.
2. To learn various staining procedures for these structures.

References

Balows, Albert, ed. *Manual of clinical microbiology,* 5th ed. American Society for Microbiology, Washington, D.C., 1991.

Gerhardt, Philip, ed. *Manual for general and molecular bacteriology.* American Society for Microbiology, Washington, D.C., 1994.

Nester et al. *Microbiology: A human perspective.* Chapter 3, 51–60.

Capsule Stain

Materials

Cultures
  Klebsiella pneumoniae or other organism with a capsule growing on a slant
India ink

Procedure

1. Make a suspension of the organism in a drop of water on a clean slide.
2. Put a drop of India ink next to it. ~~Few minutes~~
3. Carefully lower a coverslip over the two drops so that they mix together. There should be a gradient in the concentration of the ink.
4. Examine under the microscope and find a field where you can see the cells surrounded by a halo in a black background.
5. Drop slides in a beaker or can of boiling water and boil for a few minutes before cleaning. This is necessary because the bacteria are not killed in the staining process.
6. Record results.

Endospore Stain

Materials

Culture
  Bacillus cereus on nutrient agar slant after three or four days incubation at 30° C
Malachite green in staining bottles
Safranin in staining bottles
Metal or glass staining bars
Beaker or can

Figure 5.3 (a–e) Procedure for staining endospores. From John P. Harley and Lansing M. Prescott, *Laboratory Exercises in Microbiology*, 2d ed. Copyright © 1993 Wm. C. Brown Communications, Inc., Dubuque, Iowa. All Rights Reserved. Reprinted by permission.

(a) Apply malachite green to saturate paper and steam for 5 minutes

(b) Remove paper, cool, and rinse with water for 30 seconds

(c) Counterstain with safranin for 30 seconds

(d) Rinse with water for 30 seconds

(e) Blot dry with a paper towel

Procedure

1. Prepare a smear on a clean slide and heat-fix.
2. Add about an inch of water to a beaker and bring it to a boil.
3. Place two short staining bars over the beaker and place a slide on them.
4. Tear a piece of paper towel a little smaller than the slide and lay on top of the smear. The paper prevents the dye from running off the slide.
5. Flood the slide with malachite green and steam for 5 minutes. Continue to add stain to prevent the dye from drying on the slide (figure 5.3).
6. Decolorize with water for about 30 seconds by flooding with water or holding under gently running tap water. The vegetative cells (dividing cells) will lose the dye, but the endospores will retain the dye.

Figure 5.4 Appearance of endospores stained with spore stain and Gram stain. Note: The *Bacillus* frequently lose their ability to stain gram positive.

Spore stain of *Bacillus* with malachite green

Vegetative cells (pink)   Sporulated cells

Endospores (green)

Gram stain of *Bacillus*

Vegetative cells (purple)   Sporulated cells

Gram + rods   Endospores not stained

7. Counterstain with safranin for about 30 seconds and then wash with tap water for 30 seconds. Blot dry carefully.
8. Observe with the oil-immersion lens. The endospores will appear green and the vegetative cells will appear pink. Sometimes the endospore will still be seen within the cell, and its shape and appearance can be helpful in identifying the organism. In other cultures, the endospores may be free because the cells around them have disintegrated (figure 5.4).
9. Record results.

Note: When bacteria containing endospores are Gram stained the endospores do not stain and the cells appear to have holes in them. (See figure 5.4.)

Storage Granules Stain

Many organisms can store materials that are abundant in their environment for use in the future. For example, phosphate can be stored as volutin or metachromatic granules. When organisms containing these granules are stained with methylene blue, the phosphate granules are stained a darker reddish blue.

Materials

Cultures
    Spirillum grown in nutrient broth
Methylene blue in staining bottles

Procedure

1. Prepare a smear from the broth. It might be helpful to remove the organisms from the bottom of the tube with a capillary pipet. Place a drop on a clean slide. Dry and heat-fix.
2. Flood the slide with methylene blue for about 20–30 seconds.
3. Wash with tap water and blot dry.
4. Observe with the oil-immersion lens. The metachromatic granules should appear as dark reddish-blue bodies within the cells.
5. Record results.

Flagellar Stain

There are three basic kinds of arrangement of flagella:

Polar. A single flagellum at one or both ends of the organism such as *Pseudomonas*.

Lophotrichous. A tuft of flagella at one or each end of the organism, as in *Spirillum*.

Peritrichous. The surface of the organism is covered with flagella, such as *E. coli*.

Materials

Stained demonstration slides of *Escherichia coli* and *Pseudomonas*

Procedure

1. Observe flagellar stained slides of several organisms and note the pattern of flagella. It is difficult to perform this staining procedure, so prestained slides are recommended.
2. Record results.

pink = gram +   purple = gram −

NOTES:

Unknown A   (G+ cocci)   Staphylococcus aureus

          (G− rod)   Enterobacter aerogenes

     B   (G+ rod)   Bacillus subtilis

          (G− rod)   Escherrichia coli

    C (G+ rod)   chromobacter violaceum (black)

       (G+ rod)   Micrococcus luteus

    D (G+ round)   Streptococcus faecalis

     (g+   Serratia Marcescens (yellow

EXERCISE 5

Laboratory Report: Multiple and Differential Stains

Results

	Gram Reaction	Arrangement (sketch)		Optional Stains		Organism	Appearance
E. coli				Acid-fast			
B. subtilis				Capsule			
S. epidermidis				Endospore			
E. faecalis				Storage granules			
M. luteus				Flagella			

Questions

1. What is the function of each one of the Gram-stain reagents?

2. Give two reasons gram-positive organisms sometimes appear gram negative.

3. What is the purpose of using a control in the Gram stain?

4. What is a capsule?

5. What are storage granules and why are they important to the cell?

6. How does an endospore appear:
   a. when Gram stained?

   b. when spore stained?

7. What is another way you could determine whether an organism was motile besides observing a flagellar stain?

8. Why can't you Gram stain an acid-fast organism?

selective agars — inhibits certain bacteria, promotes others.

differential agars — help to be able to distinguish between different types of bacteria.

aseptic technique — 1) prevents contamination of culture
2) " " " environment

what it is
initials
316 - 1

1) finger   cold - 1 large white raised (gelatin mold), 3 medium white
             g- both rod/cocci   g+ clusters cocci

          warm — 20 small white; 2 medium; 1 beige, large, dull; 1 yellow
                            ↓   medium
                           micrococcus   shiny
                           tetrads   g+
2) shoe   cold - 9 medium white colonies   g+   staphylococcus

          warm — none

3) hair   cold — several small, white

          warm — big, white, fuzzy, circular; 1 medium smooth yellow, several
            small white

4) saliva   cold   small, white circular colonies ~30

          warm   "   "   "   " ~200

BS capsules spores

INTRODUCTION to Microbial Growth

Figure I.3.1 (a–e) Diagram of different media in different forms. From John P. Harley and Lansing M. Prescott, *Laboratory Exercises in Microbiology,* 2d ed. Copyright © 1993 Wm. C. Brown Communications, Inc., Dubuque, Iowa. All Rights Reserved. Reprinted by permission.

tryptacase soy (TSA)

(a) Broth tube
(b) Agar slant (front view)
(c) Agar slant (side view)
(d) Agar deep tube
(e) Agar plate (Petri plate)

Liquid — Agar — Agar

In order to study microorganisms, you must be able to grow or culture them. One bacterium is too small to do anything that can be easily measured, but a whole population of bacteria produce an effect big enough to be readily seen or counted. It is important that the population (culture) contain just one kind of organism. Such a culture is called a pure culture and is defined as a population of bacteria that have all grown from a single cell.

It might appear to be a very difficult problem to separate out one single bacterium from the millions of others and then permit it alone to form its own colony. Fortunately, there is a simple technique called the streak-plate method, which spreads individual bacteria on an agar plate. Colonies that grow from the widely separated bacteria are far enough apart that they can be easily transferred and studied further.

Organisms in the laboratory are frequently grown either in a broth culture or on a solid agar medium. A broth culture is useful for growing large numbers of organisms. Agar medium is used in a Petri dish when a large surface area is important, as in a streak plate. On the other hand, agar medium in tubes (called slants) is useful for storage because the small surface area is not as easily contaminated and the tubes do not dry out as fast as plates. You will be able to practice using media in all these forms (figure I.3.1).

Another important skill is the ability to prevent other bacteria from growing in the pure culture you are studying. Aseptic technique is a set of procedures designed to: 1. prevent a culture from being contaminated and 2. prevent the culture from contaminating you or your surroundings.

You will also use different kinds of media in this section. Most media are formulated so that they will support the maximum growth of various organisms, but some media have been designed to permit the growth of desired organisms and inhibit others (selective). Still other media have been formulated to change color or in some other way distinguish one colony from another (differential). These media can be very useful when trying to identify an organism.

It is also important to know how to count bacteria. You will have an opportunity to learn about several techniques for this.

In the next set of exercises, you will learn how to isolate a pure culture, to use aseptic technique, and to grow and count microorganisms. You will also be introduced to various kinds of media that are formulated for different purposes.

In these exercises, no pathogenic organisms are used, but it is very important to treat these cultures as if they were harmful because you will be able to then work safely with actual pathogens. Also, almost any organism may cause disease if there are large numbers in the wrong place.

Jock, when ye hae naething else to do, ye may be ay sticking in a tree; it will be growing Jock, when ye're sleeping.

The Heart of Midlothian [1818]
Sir Walter Scott

EXERCISE 6

Pure Culture and Aseptic Technique

Getting Started

Aseptic Technique

The two goals of aseptic (or sterile) technique are to prevent contamination of your culture with organisms from the environment and to prevent the culture from contaminating you or others.

In this exercise, you will transfer sterile broth back and forth from one tube to another using aseptic technique. The goal is to do it in such a way that you will not permit any organism in the environment from entering the tubes. You will be using both a sterile pipet and a flamed loop on the same set of broths. After you have practiced aseptically transferring the broth, you will incubate the broth tubes for a few days to determine if they are still sterile. If you used good technique, the broth will still be clear; if organisms were able to enter from the environment, the broth will be cloudy from the bacterial growth.

When you can successfully transfer sterile broth aseptically without contamination, you can use the same technique to transfer a pure culture without contaminating it or the environment.

Definitions

Aseptic. Free of contamination.

Incubate. Store cultures under conditions suitable for growth, often in an incubator.

Sterile. Aseptic; free of viable bacteria or viruses.

Objectives

1. To learn aseptic technique procedures and their importance.
2. To learn to isolate colonies using a streak-plate technique.

References

Nester et al. *Microbiology: A human perspective*. Chapter 4, 89–90.

Broth-to-Broth Transfer with a Wire Loop

> ### Materials
>
> Per student
> Tubes of TS broth, 2
> Inoculating loop

Procedure

1. Always label tubes before adding anything to them. In this exercise, you will be transferring sterile broth from one tube to another, so that both tubes will have the same label; however, in general, labeling tubes before inoculation prevents mistakes.
2. Grip the loop as you would a pencil and flame the wire portion red hot. Hold it at an angle so that you will not burn your hand (figure 6.1).
3. After the loop has cooled for a few seconds, pick up a tube in the other hand and remove the cap of the tube with the little finger (or the fourth and little fingers) of the hand holding the loop.
4. Flame the mouth of the tube by passing it through a Bunsen burner flame and then use the sterile loop to obtain a loopful of liquid from the tube. Flame the mouth of the tube and replace the cap. If you have trouble picking up a loopful of material, check to be sure that your loop is a complete circle without a gap.
5. Set down the first tube and pick up the second tube. Remove the cap, flame it, and deposit a loopful of material into the liquid

Figure 6.1 Aseptic technique for removing a loopful of broth culture. (*a*) Hold the culture tube in your left hand and the loop in your right hand (reverse if you are left-handed). Flame the loop to sterilize it. (*b*) Remove the culture tube plug or cap, and flame the mouth of the culture tube. (*c*) Insert the sterile loop into the culture tube. (*d*) Remove the loopful of inoculum, and flame the mouth of the culture tube again. (*e*) Replace the culture tube plug or cap. Place the culture tube in a test-tube rack. (*f*) Reflame the loop.

Figure 6.2 Various devices for filling pipets.

Small rubber bulb

Bulb and valve pipet filler

Pi-Pump® pipettors
Blue—for pipets up to 2 ml
Green—for pipets up to 10 ml

Automatic pipet aid

6. When learning aseptic technique, it is better to hold one tube at a time; later, you will be able to hold two or three tubes at the same time.

Transferring Broth with a Pipet

Note: Sterile pipets are used when it is necessary to transfer known amounts of material. Some laboratories use plastic disposable pipets and others use reusable glass pipets. Be sure to follow the instructor's directions for proper disposal after use (never put a used pipet on your bench top). Mouth pipetting is dangerous and is not permitted. A variety of bulbs or devices are used to draw the liquid up into the pipet and your laboratory instructor will demonstrate their use (figure 6.2).

The same broth tubes used for practice with the loop can be used to practice pipetting the broth back and forth.

Materials

TS broth tubes from previous procedure

1-ml pipet

Bulb or other device to fit on end of the pipet

of the second tube. Withdraw the loop, flame the tube, and then replace the cap. Be sure to flame the loop before setting it on the bench (your loop would normally be contaminated with the bacteria you were inoculating). It is usually convenient to rest the hot loop on the edge of the Bunsen burner.

Procedure

First Session

1. Open a sterile pipet at the top, insert a bulb on the end, then carefully remove the pipet from the package or canister without touching the tip. Grip the pipet as you would a pencil. The pipet is plugged with cotton to filter the air going into it. Discard the pipet if liquid inadvertently wets the plug—air will no longer enter the pipet and the measured liquid will not flow out. Notify your instructor if the bulb is contaminated.
2. Pick up a tube with your other hand and remove the cap with the little finger of the hand holding the pipet. Flame the tube. Expel air from the rubber bulb and insert the pipet tip into the liquid. Draw the liquid up to the desired amount, remove the pipet, flame the tube, replace the cap, and then put the tube back in the rack.
3. Pick up the second tube and repeat the steps used with the first tube except that the liquid is expelled into the tube.
4. Repeat the above steps with the same tubes until you feel comfortable with the procedure.
5. Dispose of the pipet as directed.
6. Incubate the tubes until the next period at 37° C.

Second Session

1. Observe the tubes of broth for turbidity. If they are cloudy, organisms contaminated the broth during your practice and grew during incubation. With a little more practice, you will have better technique. If the broths are clear, there was probably no contamination and you transferred the broth without permitting the entry of any organisms into the tubes.
2. Record results.

Streak-Plate Technique

Materials

Per student

  Trypticase soy agar (TSA) plates, 2
  Cultures
    Broth culture containing a mixture of two organisms such as *Micrococcus* and *Staphylococcus*

Procedure

First Session

1. Label the agar plate on the bottom with your name and date.
2. Divide the plate into three sections with a "T" as diagrammed (figure 6.3).
3. Sterilize the loop in the flame by heating the whole length of the wire red hot. Hold it at an angle so you do not heat the handle or roast your hand.
4. Gently shake the culture to be sure both organisms are suspended. Aseptically remove a loopful of the culture and holding the loop as you would a pencil, spread the bacteria on section 1 of the plate by streaking back and forth. The more streaks, the better chance of an isolated colony. As you work, partially cover the Petri dish with the cover to minimize organisms falling on the plate from the air. Use a gliding motion and avoid digging into the agar. Don't press the loop into the surface. If your loop is not smooth or does not form a complete circle, it can gouge the agar and colonies will run together. Note that you can see the streak marks if you look carefully at the surface of the plate.

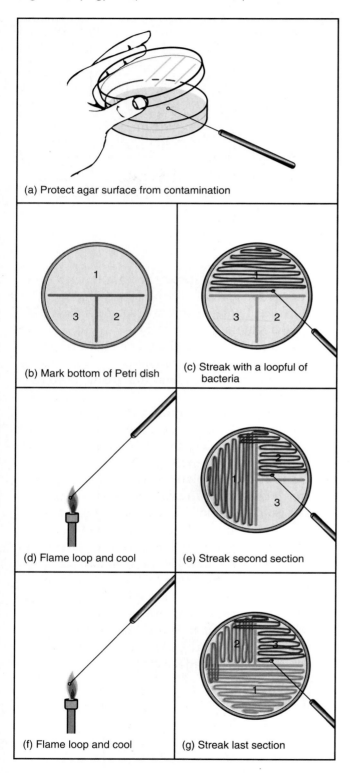

Figure 6.3 (*a–g*) Preparation of a streak plate.

(a) Protect agar surface from contamination

(b) Mark bottom of Petri dish

(c) Streak with a loopful of bacteria

(d) Flame loop and cool

(e) Streak second section

(f) Flame loop and cool

(g) Streak last section

5. Burn off all the bacteria from your loop by heating it red hot. This is very important because it eliminates the bacteria on your loop. Wait a few seconds to be sure the loop is cool.

6. Without going into the broth again, streak section 2 (see figure 6.3) of the Petri plate. Go into section 1 with about three streaks and spread by filling section 2 with closely spaced streaks.

7. Again heat the loop red hot. Go into section 2 with about three streaks and spread by filling section 3 with streaks.

8. Heat loop red hot before placing it on the bench top. Usually you can rest it on some part of the Bunsen burner so that it can cool without burning anything.

9. Repeat the procedure with a second plate for additional practice.

10. Incubate the plates in the 37° C incubator.

Second Session

Observe your streak plates and record results.

danot re

C = Chromobacterivolaceum - blac

D - Serratia Marcescens - pin

B - gram -

A - gram +

EXERCISE

6

Laboratory Report: Pure Culture and Aseptic Technique

Results

B

	Clear	Turbid
Tube 1		
Tube 2		

	Number of Colonies Isolated
Streak plate 1	19 big, puffy
Streak plate 2	

Questions

1. What is the definition of a pure culture?

2. Why is sterile technique important?

3. Why is it important to avoid digging into the agar with the loop?

4. Is there anything you can do to improve your streak-plate technique?

EXERCISE 7

Defined, Undefined, Selective, and Differential Media

Getting Started

Microbiologists have developed several different types of culture media for a variety of different purposes. If a medium is totally made up of specific amounts of chemicals it is called a **defined medium.** If, however, it contains such mixtures as yeast extract or enzymatic digests of protein, it is termed an **undefined medium,** or complex medium, because the exact amount and kinds of large organic molecules are not known.

Undefined media are frequently called rich media. These media tend to support the growth of bacteria better because they contain more preformed nutrients, and the organisms do not have to use energy and materials to synthesize the compounds supplied in the medium. Many bacteria can grow only on this kind of medium because they cannot synthesize all the necessary components needed for growth and must be provided with preformed amino acids, vitamins, and other growth factors. Some organisms require only one or two vitamins or amino acids while other bacteria need many complex growth factors, and are termed fastidious.

Two other valuable types of media are **selective** and **differential media.** Frequently, it is important to isolate one organism in a mixture of bacteria. Normal flora can contaminate the culture; for example, a wound culture may be mixed with large amounts of *Staphylococcus* normally found on the skin. It can be difficult to isolate or even detect the pathogenic organism among all the nonpathogenic organisms present. Selective media have been designed to permit some bacteria to grow but not others, so that certain bacteria can be isolated even if they constitute only a small percentage of the population.

Differential media are also useful for isolating and identifying bacteria. By observing the appearance of colonies growing on this agar, it is possible to determine characteristics such as whether or not they can ferment certain sugars.

In this exercise, you will have an opportunity to observe the growth of organisms on three different media.

Glucose Salts Agar This is a simple defined medium. Only organisms that can make all their cellular components from glucose and inorganic salts are able to grow on it.

Trypticase Soy Agar This is a rich, undefined medium made from an enzymatic digest of protein and soy product. Organisms that require vitamins or other growth factors are able to grow on it.

EMB (Eosin Methylene Blue) Agar This is a selective medium permitting the growth of gram-negative enteric rods and inhibiting the growth of gram-positive bacteria. In addition, the medium is also differential because it contains the sugar lactose. Organisms that can ferment lactose produce purple colonies and those that cannot, produce white or very light pink colonies. The colonies of *E. coli*, a lactose fermenter, are dark purple. They also give the medium a distinctive metallic green sheen caused by the large amounts of acid produced. The colonies of *Enterobacter*, also a lactose fermenter, usually are more mucoid with purple centers. (Mucoid colonies have a slimy appearance.)

Definitions

Defined medium. A synthetic medium composed of inorganic salts and usually a carbon source such as glucose.

Differential medium. Medium permitting certain organisms to be distinguished from others by the appearance of their colonies.

Rich (or **enriched**) **medium.** A medium containing many growth factors. Usually it is an undefined medium made from meat, plant, or yeast extracts.

Figure 7.1 A diagram of the labeled media plates.

Trypticase soy   Glucose mineral   Eosin methylene blue (EMB)

Selective medium. Medium formulated to permit the growth of certain bacteria but not others.

Undefined medium. A complex medium in which the exact amounts of components and their composition are unknown because it is made of extracts or enzymatic digests of meat, plants, or yeast.

Objectives

1. To compare the growth of organisms on a defined and a rich medium.
2. To compare the growth of organisms on a selective and a differential medium.
3. To understand the relationship between the growth of an organism and the composition of the medium.

References

Nester et al. *Microbiology: A human perspective*. Chapter 4, 98–100.

Procedure

First Session

1. With a marking pen, divide the bottom of the Petri plates into quadrants. Label the plates with your name and date. Label each quadrant with the organism as shown in figure 7.1.
2. Inoculate each quadrant of the plate with a loopful of the culture in a wavy line.
3. Invert and incubate at 37° C for 48 hours.

Second Session

1. Observe and compare the growth on the three plates.
2. Record the results.

Materials

Cultures growing in trypticase soy broth
  Escherichia coli
  Staphylococcus epidermidis
  Pseudomonas aeruginosa
  Enterobacter aerogenes

Media per team

Trypticase soy (TS) agar plate

Glucose mineral salts agar plate

EMB (Eosin Methylene Blue) agar plate

EXERCISE 7

Laboratory Report: Defined, Undefined, Selective, and Differential Media

Results

	Glucose salts	TSA	EMB
1 E. coli	$\frac{1}{2}$ +	+ + +	+ + + *cool*
2 Staphylococcus epidermidis	$\frac{1}{2}$ +	+ + +	+
3 Pseudomonas aeruginosa	+ + +	+ + +	++ +
4 Enterobacter aerogenes	+	+ + +	+ + +

Indicate the amount of growth:
- 0=no growth
- +=slight growth
- ++=good growth
- +++=excellent growth
- Lac −=no lactose fermentation on EMB
- Lac +=lactose fermentation on EMB

Questions

1. Which organisms could not grow on the glucose salts medium? Which organisms could grow on it?

2. Which organisms do not require any growth factors?

3. Of the organisms that could grow on both TS agar and glucose salts, did some organisms grow better on the TS agar than the glucose salts? Can you propose a reason?

4. Which organisms could grow on the EMB agar?

5. Which organisms could ferment lactose?

6. Could you differentiate *E. coli* from other organisms growing on EMB?

7. In general, EMB selects for what kind of organisms?

8. What kinds of organisms does EMB differentiate?

EXERCISE

8

Quantification of Microorganisms

Getting Started

It is frequently important to count bacteria. For example, you may want to know the number of bacteria in a sample of raw chicken or the number of bacteria per ml of water in a swimming pool. Special techniques have been devised to enumerate bacteria, each with advantages and disadvantages. Three common methods are discussed.

Plate Count This method is based on the premise that each viable bacterium will produce a colony when growing on an agar plate. A sample of the material to be counted is suspended in liquid and placed in an empty Petri plate. Melted, cooled agar is then added to the plate and mixed with the inoculum. After incubation, each organism produces a colony in the agar that can then be counted. The plate count is used very frequently but it has advantages and disadvantages that should be considered prior to use. Some of these are discussed next.

1. Bacteria are usually present in very large numbers—an overnight broth culture of *E. coli* can easily contain one billion cells/ml. However, the maximum number of colonies that can be accurately counted on a plate is usually set at 300. Therefore, most samples must be diluted to low enough numbers that the plates will have distinct colonies that can be counted. Since it is usually not possible to know exactly how dilute to make a sample to obtain a countable plate, several different dilutions must be plated.
2. Some bacteria tend to stick together; therefore, sometimes two or more bacteria will give rise to one colony. This gives results of a lower number than are actually present.
3. If a sample has many different kinds of bacteria, it is not possible to have a medium or conditions that will support all their various necessary growth conditions. Soil, for example, may contain organisms that will not grow unless the temperature is above 50° C;

in contrast, other organisms are inhibited at these temperatures. These problems must be considered when a sample of mixed bacteria is enumerated.

4. It can take at least 24 hours to obtain the results of a plate count.
5. The plate count method does have two advantages over other methods, however. Only viable organisms are counted, which are the ones usually considered important. Also, samples with small numbers can be counted, which would have insufficient numbers for other methods.

Direct Count In this method of counting organisms, a suspension of bacteria is placed on a slide that has been ruled into squares and is designed to hold a specific volume of liquid. By counting the bacteria that appear on the grid areas, the number of organisms in the sample can be calculated. It is a much faster test than the plate count, but it does have several drawbacks. First, there must be about 1×10^7 organisms/ml before there are enough to be seen, and second, viable and nonviable organisms appear the same under a microscope.

Turbidometric In this method, a spectrophotometer is used to measure the turbidity or optical density (O.D.) of bacteria in a broth. The more bacteria, the cloudier the broth and the higher the optical density. In this method, you must first correlate plate counts with optical density readings. This must be done with each strain of bacteria because organisms are different sizes. For instance, an optical density reading of 0.2 for a broth culture of one *E. coli* strain is equal to 1×10^8 cells/ml. The same number of another organism would have a different optical density. Once the correlation between O.D. and plate counts has been determined, the correlation can be used as an extremely convenient method of determining numbers of organisms. This method is used to determine the generation time in exercise 10. Other methods of enumerating bacteria are discussed in exercise 34, which is on water analysis.

The plate count method is used in this exercise to count the number of organisms in two broth cultures: one turbid (sample A) and the other with no visible turbidity (sample B). There are two methods of preparing plate counts: pour plates and spread plates. In the pour plate method (as just described), a sample is mixed with melted agar in a Petri plate and colonies appearing in and on the agar are counted. In the spread plate method, a small sample is placed on the surface of the agar plate and spread with a bent glass rod so that all the colonies appear on the surface of the plate. The spread plate method is used in exercise 17, Bacterial Conjugation.

Definitions

Optical Density. (O.D.) A measure of the amount of turbidity. Frequently called absorbance.

Serial dilution. Preparing a dilution in steps instead of one dilution.

Turbid. Cloudy.

Viable (bacteria). Capable of growing and dividing.

Objectives

1. To enumerate bacteria using the plate count method.
2. To understand the use of dilutions.
3. To gain insight into the number of organisms that can be present in a clear liquid and a turbid liquid.

References

Nester et al. *Microbiology: A human perspective.* Chapter 4, 90–92.

Note: When serially transferring concentrated suspensions to less concentrated suspensions, a new pipet must be used for each transfer to prevent car-

---

ryover of organisms. (In this exercise, the instructor may choose to use each pipet twice to conserve materials.) However, when going from less concentrated to more concentrated suspensions, the same pipet may be used with no significant error because only a few organisms will be added to the much larger sample.

Procedure for Team A

First Session

1. Label all water blanks with the dilution, as shown in figure 8.1.
2. Melt 4 TS agar deeps and hold at 50° C. It is very important not to let the deeps cool much lower than 50° C because the agar will harden and will have to be heated to boiling (100° C) before it will melt again (figure 8.2).
3. Make serial dilutions of the bacterial suspension.
   a. Mix the bacterial suspension by rotating between the hands and transfer 1.0 ml of the suspension to the 99-ml water blank labeled 10^{-2}. Discard the pipet.
   b. Mix well and transfer 0.1 ml of the 10^{-2} dilution to the 9.9-ml water blank labeled 10^{-4}. Discard pipet.
   c. Mix well and transfer 0.1 ml of the 10^{-4} dilution to the 9.9-ml tube labeled 10^{-6}. Discard pipet.

Figure 8.1 Dilution scheme for Team A.

Team A

1.0 ml · Sample · 0.1 ml · 0.1 ml · 1.0 ml · 1.0 ml · 1.0 ml

99 ml 10^{-2} · 9.9 ml 10^{-4} · 9.9 ml 10^{-6} · 9.0 ml 10^{-7} · 9.0 ml 10^{-8} · 9.0 ml 10^{-9}

Figure 8.2 (a–f) Melting and pouring agar deeps.

(a) Boiling water bath

Thermometer

(b) 50°C water bath

(c) Wipe tube with paper towel

(d) Flame the top of the tube after removing cap

(e) Pour the agar into the Petri dish bottom

Agar

(f) After cooling, store in an inverted position

d. Mix and transfer 1.0 ml of the 10^{-6} dilution to the 9.0-ml tube labeled 10^{-7}. (Note change from 0.1 ml to 1.0 ml.) Discard pipet.
e. Mix and transfer 1.0 ml of the 10^{-7} dilution to the 9.0-ml tube labeled 10^{-8}. Discard pipet.
f. Mix and transfer 1.0 ml of the 10^{-8} dilution to the 9.0-ml tube labeled 10^{-9}. Discard pipet.

4. Place samples of the dilutions 10^{-9}, 10^{-8}, 10^{-7}, and 10^{-6} into sterile labeled Petri plates as follows:
   a. Mix the 10^{-9} dilution and transfer 1.0 ml into a sterile Petri plate labeled 10^{-9}. Add a tube of melted agar (wipe off the outside of the tube before pouring) and swirl gently by moving the plate in a figure eight pattern on the bench. Do not discard pipet. (See figure 8.2.)
   b. Mix the 10^{-8} dilution and with the same pipet transfer 1.0 ml to the Petri plate labeled 10^{-8}. Add melted agar and mix.
   c. Mix the 10^{-7} dilution again and using the same pipet transfer 1.0 ml to the Petri plate labeled 10^{-7}. Add melted agar and mix.
   d. Mix the 10^{-6} dilution and using the same pipet transfer 1.0 ml to the Petri plate labeled 10^{-6}. Add melted agar and mix. Discard the pipet.
5. Invert the plates after you are sure the agar has hardened (about 5 minutes), and incubate at 37° C.

Figure 8.3 Dilution scheme for Team B.

Team B

Sample

1.0 ml → 0.1 ml → 1.0 ml → 1.0 ml → 1.0 ml

99 ml
10^{-2}

9.9 ml
10^{-4}

9.0 ml
10^{-5}

9.0 ml
10^{-6}

9.0 ml
10^{-7}

Materials

For teams using Sample B (a nonturbid suspension)

99-ml water blank, 1

9.9-ml water blank, 1

9.0-ml water blanks, 4

1-ml pipets, 5

TS agar deeps, 4

Sterile Petri dishes, 4

Suspension B: a nonturbid suspension of bacteria

Procedure for Team B

1. Label all water blanks and plates with the dilution (figure 8.3).
2. Melt 4 TS agar deeps and hold at 50° C. It is very important not to let the agar deeps cool much lower than 50° C because the agar will harden and will have to be heated to boiling (100° C) before it will melt again.
3. Make serial dilutions of the bacterial suspension.
   a. Mix the bacterial suspension and transfer 1.0 ml of the suspension to the 99-ml water blank labeled 10^{-2}. Discard the pipet.
   b. Mix well and transfer 0.1 ml of the 10^{-2} dilution to the 9.9-ml water blank labeled 10^{-4}. (Note that this is the only time you will use a 0.1 ml-sample.) Discard pipet.
   c. Mix well and transfer 1.0 ml of the 10^{-4} dilution to the 9.0-ml tube labeled 10^{-5}. Discard the pipet.
   d. Mix and transfer 1.0 ml of the 10^{-5} dilution to the 9.0-ml tube labeled 10^{-6}. Discard the pipet.
   e. Mix and transfer 1.0 ml of the 10^{-6} dilution to the 9.0-ml tube labeled 10^{-7}. Discard the pipet.
4. Place samples of the dilutions 10^{-7}, 10^{-6}, 10^{-5}, and 10^{-4} into the pour plates as follows.
   a. Mix the 10^{-7} dilution tube and remove 1 ml to the Petri plate labeled 10^{-7}. Pour the melted, cooled agar into the plate (wipe off any water on the outside of the tube before pouring the agar). (See figure 8.2.) Mix by gently moving the plate in a figure eight on the bench. Do not discard the pipet.
   b. Repeat the procedure for the 10^{-6}, 10^{-5}, and 10^{-4} dilutions using the same pipet.
5. After the agar has hardened (about 5 minutes) invert the plates and incubate at 37° C.

Second Session for Both A and B Teams

1. Count the colonies in the agar plates. Use a marking pen on the bottom of the plate to dot the colonies as you count them. Colonies growing in the agar tend to be lens-shaped and smaller than those growing on the surface but all are counted equally. If there are more than 300 colonies on the plate, label it TNTC—too numerous to count.

2. Choose the plate that has between 30 and 300 colonies (less than 30 gives results with a high sample error). Calculate the number of organisms/ml using the following formula:

   number of organisms on the plate $\times$ 1/sample volume $\times$ 1/dilution = the number of organisms/ml in the original suspension.

Note: In this exercise the volume of all the samples is 1.0 ml.

3. Record your results. Post your results on the blackboard so that average numbers of organisms/ml for Suspension A and Suspension B can be calculated.

ethereal

NOTES:

aerobes = O_2 - respiration

microaerophiles - small amounts of O_2 (too much lethal)

obligate anaerobes - fermentative enzymes
(no O_2 - no tolerance)

aerotolerant - don't use O_2 but tolerable

facultative anaerobes - prefer O_2 - can grow either way

aerobes   m.c.   o.a.   f.a.

Gas Pack Jar

palladium catalyst

blue - O_2 white - no O_2

Bacterial Growth Curve
Generation Time
TSB

spectrometer
600nm

lag phase - no growth
logarimithic growth (log phase)

stationary phase
decline phase

Name _____ Date _____ Section _____

EXERCISE 8

Laboratory Report: Quantification of Microorganisms

Results

1. Which plate (dilution) had between 30 and 300 colonies?
2. How many colonies did you count?
3. How many organisms/ml were in the original suspension? (Show calculations)

	Solution A	Solution B
Appearance of broth		
Number of organisms/ml (class average)		
Number of organisms/ml (your data)		

Questions

1. From these results, about how many organisms/ml can be in a cloudy broth? (Show calculations.)

2. From these results, about how many organisms/ml can be in a clear broth without showing any sign of turbidity? (Show calculations.)

3. What are two sources of error in this procedure?

4. If you serially dilute a sample with three 1:10 dilutions, what is the final dilution of the last tube?

5. If you add 1.0 ml to 99 ml of water, what is the dilution of the sample?

Understanding Dilutions

1. To make a dilution use the following formula:

   sample/(diluent + sample) = the dilution

Example 1 How much is a sample diluted if 1 ml is added to 9.0 ml of water (the water is sometimes called a diluent)?

   1/(1 + 9) = 1/10 (also expressed 10^{-1})

Example 2 How much is a sample diluted if 0.1 ml is added to 9.9 ml of water?

   $$0.1/(0.1 + 9.9) = 0.1/10$$
   $$= 1:100 \text{ or } 10^{-2}$$

2. When a sample is serially diluted, multiply each dilution together for the final dilution.
   The final dilution in tube B is 1:100 or 10^{-2}.

3. To calculate the numbers of organisms in the original suspension use the formula:

   The number of organisms/ml in the original sample = number of colonies on plate × 1/volume of sample × 1/dilution

1 ml   1 ml

9 ml   9 ml
Tube A   Tube B
10^{-1} 10^{-2}
(1/10)   (1/100)

Example 3 Suppose you counted 120 organisms on a plate diluted 10^{-2}. The sample size was 0.1 ml.

   Solution: 120 (number of organisms on plate) × 1/0.1 × $1/10^{-2} = 120 \times 10 \times 100$
   $= 120 \times 10^3 \text{ or } 1.2 \times 10^5$ organisms/ml

Example 4 Suppose you counted 73 colonies on the plate marked 10^{-6}. If the sample size is 1.0 ml then

   $73 \times 1/1.0 \times 1/10^{-6} = 73 \times 1 \times 10^6 = 73 \times 10^6 \text{ or } 7.3 \times 10^7$
   organisms/ml in the original suspension

(It is important to label the answer "per ml.")

INTRODUCTION to the Environment and Microbial Growth

An organism cannot grow and divide unless it is in a favorable environment. Environmental factors include temperature, availability of nutrients, moisture, oxygen, salinity, and presence of toxic products.

Each bacterial species has its own particular set of optimal conditions that allows maximum growth. These conditions probably reflect the environment in which the organism grows and competes with other organisms. In the next two exercises, we examine the effects of temperature and atmosphere (oxygen) on the growth of bacteria. (Another environmental factor affecting growth is osmotic pressure, which is examined in exercise 13.)

The dream of a bacterium is to become two bacteria.

F. Jacob

---

streptococcus

1 2 3 4 5 6 7 8 9 10 11 12 13

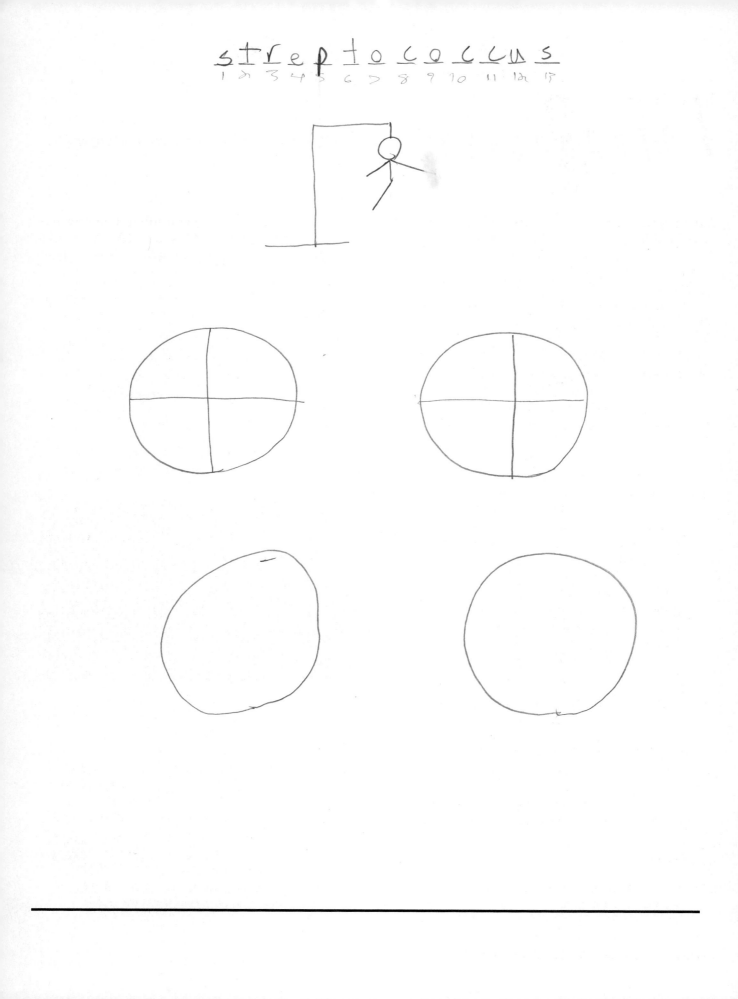

EXERCISE 9

Aerobic and Anaerobic Growth

Getting Started

All the animals we are familiar with, including humans, have an absolute requirement for oxygen. It seems rather surprising then that there are groups of organisms that cannot grow or are even killed in the presence of oxygen. Still other kinds of organisms can grow either with or without oxygen. These three groups are classified as follows:

1. *Obligate aerobes* Organisms that have an absolute requirement for oxygen. *Micrococcus*, a member of the skin flora, and *Pseudomonas*, a soil organism (and occasional pathogen), are obligate aerobes.

2. *Obligate anaerobes* Organisms that cannot grow in the presence of oxygen. There are, however, varying degrees of sensitivity to oxygen. Some bacteria such as the methanogens that produce methane gas in swamps are killed by a few molecules of O_2 while others, such as *Clostridium*, can survive in O_2 but cannot grow until conditions become anaerobic.

3. *Facultative anaerobes* Organisms that can grow either in the presence or absence of oxygen. (Usually they are simply called facultative.) *Escherichia coli*, a member of the intestinal flora, is facultative.

There are also other categories such as microaerophilic organisms, which do best in reduced amounts of oxygen, and organisms that prefer more CO_2 than the amount normally found in the atmosphere. In this exercise, however, we examine the oxygen requirements of an obligate aerobe, an obligate anaerobe, and a facultative anaerobe. Try to identify them by growing each culture aerobically on a slant and anaerobically in an agar deep.

Definitions

Aerobic. In the presence of air. Air contains about 20% oxygen.

Agar deep. A test tube filled with agar almost to the top.

Anaerobic. In the absence of air.

Microaerophilic. Reduced amounts of air.

Objectives

1. To understand how microbes differ in their oxygen requirements.
2. To learn to distinguish between aerobes, anaerobes, and facultative anaerobes.

References

Nester et al. *Microbiology: A human perspective.* Chapter 4, 95–96; 100–102; figures 4.9, 4.11.

Materials

Per team

TS broth cultures labeled A, B, and C
 Escherichia coli
 Micrococcus
 Clostridium
TS + 0.5% glucose agar deeps, 3
TS agar slants, 3

Procedure

First Session

1. Put the agar deeps in a beaker, or can, and fill it with water to the height of the agar. Boil the tubes for 10 minutes. This will not only melt the agar but also drive out all the dissolved oxygen (figure 9.1).

 After the agar hardens, air will gradually diffuse into the tube so that about the top half inch of the agar will be aerobic, but the remainder of the tube will be anaerobic.

Figure 9.1 Boiling agar deeps for 10 minutes to drive out dissolved oxygen.

Figure 9.2 (*a*) Inoculating a melted agar deep. (*b*) Inoculating an agar slant.

(a) An agar deep   (b) Inoculating an agar slant

Figure 9.3 The appearance of aerobic and anaerobic growth in shake tubes.

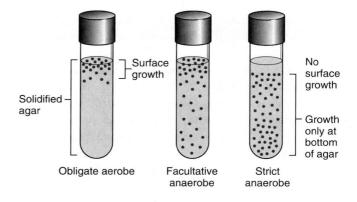

Solidified agar   Surface growth

No surface growth

Growth only at bottom of agar

Obligate aerobe   Facultative anaerobe   Strict anaerobe

2. Cool the agar in a 50° C water bath for about 10 minutes (check the temperature of the water with a thermometer). Be careful that the agar does not cool much lower than 50° C or it will solidify and you will have to boil it to melt it again. The tube will feel hot, but you will be able to hold it.
3. Label the tubes and slants.
4. Inoculate a melted agar deep with a loopful of culture A (figure 9.2) and mix by rolling between the hands. Permit the agar to harden. This technique is often called a shake tube.
5. Inoculate a slant with a loopful of culture A by placing a loopful of broth on the bottom of the slant and making a wiggly line on the surface to the top of the slant.
6. Repeat with cultures B and C.
7. Incubate all slants and deeps at least 48 hours at 25° C. Some cultures grow so vigorously at 37° C that the gas produced blows apart the agar.

Second Session

1. Observe the surface of the slants and of the deeps, and record the growth. Compare to figure 9.3.

   Note: Sometimes the anaerobes seep down and grow between the agar slant and the walls of the glass tube where conditions are anaerobic, but not on the surface of the slant, which is aerobic.

2. Which cultures were the aerobes, the facultative anaerobes, and the anaerobes?

EXERCISE

9

Laboratory Report: Aerobic and Anaerobic Growth

Results

Culture	A 1	B 2	C 3
aerobic Growth on slant	+++ white	+	+++ yellow
anaerobic	+++	+++	○
anaerobic. Appearance of growth in deep			

Identity of culture A

culture B

culture C

Questions

1. Would you expect the obligate anaerobe to grow on a slant incubated aerobically? Why?

2. Which organism would you expect to grow both in the agar deep and on the slant?

3. Which organisms could grow aerobically on slants?

4. Which of the organisms could grow throughout the agar deep and on top?

5. Why did you boil the agar deeps longer than it took to melt the agar?

6. If air can diffuse into agar and broth, how were the obligate anaerobes grown in the broth for the class? Your instructor will explain or demonstrate.

EXERCISE 10

The Effect of Incubation Temperature on Generation Time

Getting Started

Every bacterial species has an optimal temperature—the particular temperature resulting in the fastest growth. Normally, the optimal temperature for each organism reflects the temperature of its environment. Organisms associated with animals usually grow fastest at about 37° C, the average body temperature of most warm-blooded animals. Organisms can divide more slowly at temperatures below their optimum, but there is a minimum temperature below which no growth occurs. Bacteria usually are inhibited at temperatures not much higher than their optimum temperature.

The effect of temperature can be carefully measured by determining the generation time at different temperatures. Generation time, or doubling time, is the time it takes for one organism to divide into two cells; on a larger scale, it is the time required for the population of cells to double. The shorter the generation time, the faster the growth rate.

Generation time can only be measured when the cells are dividing at a constant rate. To understand when this occurs, it is necessary to study the growth curve of organisms inoculated into a fresh broth medium. If plate counts are made of the growing culture, it can be seen that the culture proceeds through the four phases of growth: Lag, Log, Stationary, and Death (figure 10.1).

In the lag phase, the cells synthesize the necessary enzymes and other cellular components needed for growth. The cells then grow as rapidly as the conditions permit in the log phase, and when there are no longer sufficient nutrients or

Figure 10.1 Growth curve showing the four phases of growth.

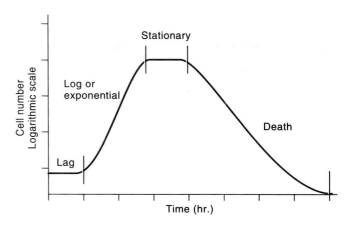

toxic product buildup, the cells go into the stationary phase. This is followed by the death phase. Only in the log phase are the cells growing at a constant maximum rate for the particular environment.

In this exercise, the generation time of *E. coli* will be compared when growing at two different temperatures. The growth of the cells can be measured with a spectrophotometer or colorimeter because the number of cells in the culture is directly proportional to the absorbance. That means that the absorbance (also called optical density) increases proportionately as the culture becomes increasingly more cloudy from the multiplication of the bacteria. Readings of the incubating cultures are taken every 20 minutes for 80 minutes. The results are then plotted, and the generation time is determined.

Definitions

Absorbance.   A measure of turbidity.

Generation time.   The time it takes for a population of cells to double.

Optical density. (O.D.)   An older, but still widely used, term for absorbance when used for measuring bacterial growth.

Spectrophotometer.   An instrument that can be used to measure the turbidity of bacterial growth.

Objectives

1. To understand the phases of a growth curve.
2. To understand the effect of temperature on generation time.
3. To learn to calculate generation time.
4. To learn to use semi-log paper.

References

Nester et al. *Microbiology: A human perspective*. Chapter 4, 104–106; figure 4.16.

Materials

Per team

Cultures
  Escherichia coli (TS broth cultures in log phase)

TS broth, 1, in a tube that can be read in a spectrophotometer or Klett. Prewarmed in a water bath.

Water bath at 30° C (to be used by half the class)

Water bath at 37° C (to be used by the other half of the class)

Spectrophotometer
Courtesy of Dr. Harold J. Benson

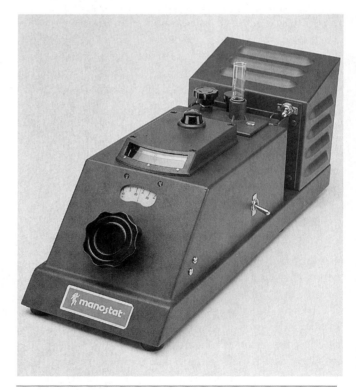

Klett Colorimeter
Courtesy of VWR Scientific Company

Procedure

1. Add 0.5 ml–1.0 ml of an *E. coli* culture growing in log phase to 5.0 ml TS broth. The dilution is not important as long as the broth is turbid enough to be read at the low end of the scale (0.1 O.D. or a Klett reading of about 50). If you start with an O.D. that is too high, your last readings will reach the part of the scale that is not accurate (an O.D. of about 0.4 or about 200 on the Klett).

2. With a wavelength of 420, set the spectrophotometer at zero with an uninoculated tube of TS broth (which is termed a blank). Your instructor will give specific directions.

3. Take a reading of the culture and record it as 0 time. Return tube to the assigned water bath as quickly as possible because cooling slows the growth of the organisms. Wipe off water and fingerprints from the tubes before taking a reading.

4. Read the O.D. of the culture about every 20 minutes for about 80 minutes. Record the exact time of the reading so the data can be plotted correctly.

5. Record your data—the time and O.D. readings—in your manual and on the blackboard.

6. Plot the data on semi-log graph paper (page 83). Semi-log paper is designed to convert numbers or data to $\log_{10}$ as they are plotted on the y axis. The same results would be obtained by plotting the $\log_{10}$ of each of the data points on regular graph paper but semi-log paper simplifies this by permitting you to plot raw data and obtain the same line. Time is plotted on the horizontal x axis. Draw a straight best-fit line through the data points. The cells are growing logarithmically, so therefore the data should generate a straight line on log paper (figure 10.2).

7. Also plot the data generated from the other temperature by averaging the class data on the blackboard.

8. Calculate the generation time for *E. coli* at each temperature. This can be done by arbitrarily selecting a point on the line and noting the O.D. Find the point on the line where this number has doubled. The time between these two points is the generation time.

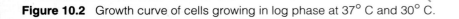

Figure 10.2 Growth curve of cells growing in log phase at 37° C and 30° C.

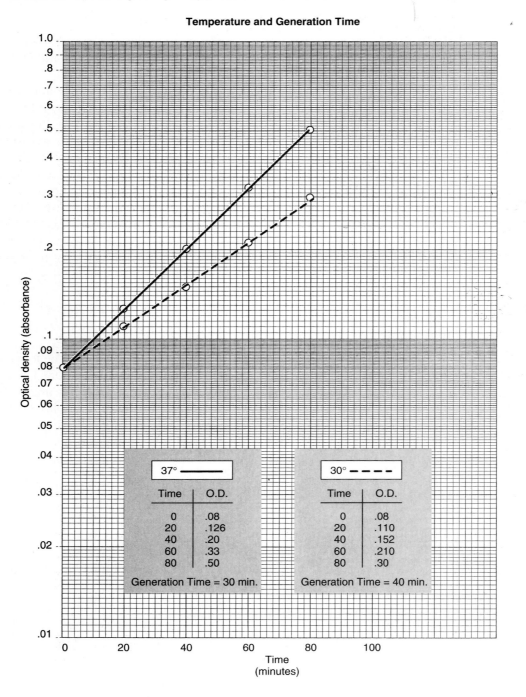

Temperature and Generation Time

37° ———	
Time	O.D.
0	.08
20	.126
40	.20
60	.33
80	.50

Generation Time = 30 min.

30° – – – –	
Time	O.D.
0	.08
20	.110
40	.152
60	.210
80	.30

Generation Time = 40 min.

EXERCISE 10

Laboratory Report: The Effect of Incubation
Temperature on Generation Time

30   37

0.03   0.03

9:08   9:08

Results

Data: Your Temperature _____		Class Average Temperature _____	
Time	Reading	Time	Reading
30 1 9:08	.03	9:28	.04
37 2 9:08	.03	9:28	.03
30 3 9:48	.05	10:08	.07
37 4 9:48	.05	10:08	.07
30 5 10:28	.08	10:48	.15
37 10:28	.12	10:48	.20
E. coli at 37° C 11:08	.25	**Generation Time** 37 11:28 .31	37 11:48
E. coli at 30° C 11:28	.17	30 11:28 .19	30 11:48

Questions

1. What is the generation time of an organism?

2. Why is it important to keep the culture at the correct incubation temperature when measuring the generation time?

3. Why is it important to use cells in log phase?

4. If the growth of two cultures were plotted on semi-log paper, one slower than the other, which would have the steeper slope?

Figure 10.3 Semi-log paper.

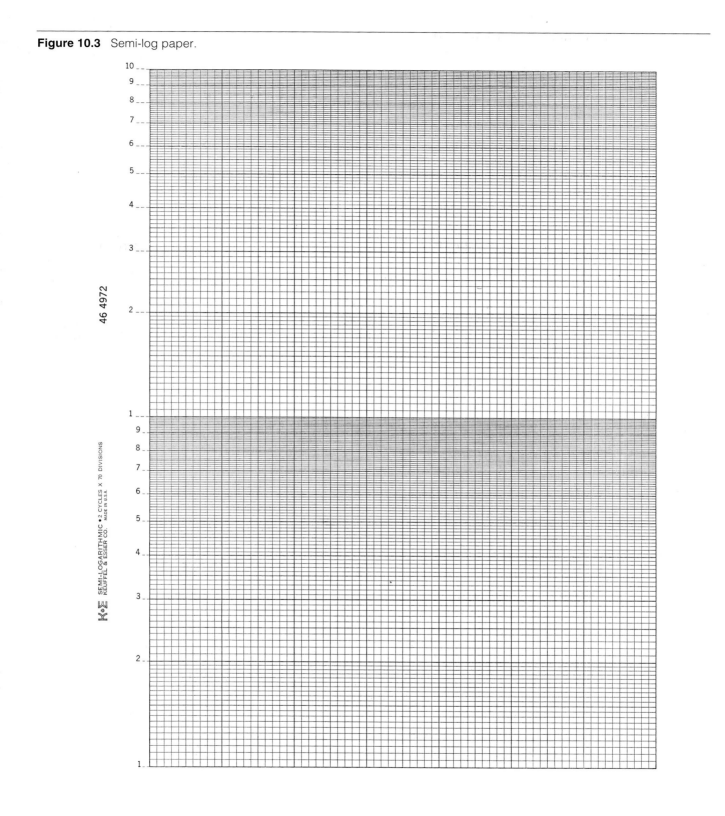

NOTES: Carbon: Autotrophs: CO_2
Heterotrophs: organic (glucose)

Nitrogen (found in proteins / amino acids) used to construct DNA, RNA

physical
temperature
  enzyme reaction time
pH

defined media - basic - agar, glucose, water
undefined (complex) - yeast + peptone extracts

MacConkey's (diff)
crystal violet - gram -
lactose, bile salts

Eosin Methylene Blue (EMB)
gram - , eosin
    L E.coli - black/gre

selective
Mannitol Salt (7.5% NaCl) Staphylococcus
  if Mannitol is used - colonies grow yellow
Sabourand agar - fungi (pH = 5.6)

PEA - gram positive

Glucose Salt Agar
  defined media

TSA
undefined agar
(extra vitamins)

Diff

blood 5% sheep
hemolysis
α   β   γ
| |
lyse complete
green clear

*I*ntroduction to Control of Microbial Growth

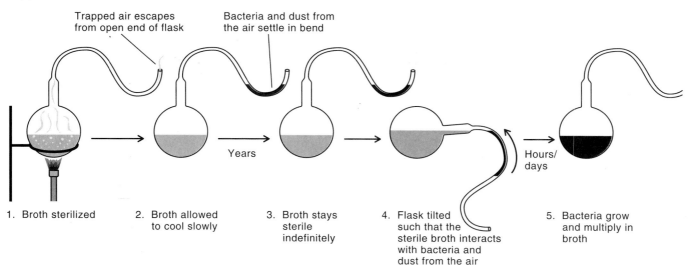

For many microbiologists, control of microbial growth means maximization of microbial growth, for example, when producing baker's yeast or in the production of antibiotics. To others such as physicians and allied members of the medical profession, control means minimization of microbial growth, for example, the use of heat and ultraviolet light to destroy microorganisms present in growth media, gloves, and clothing. It can also imply the use of **antiseptics, disinfectants,** and **antibiotics** to inhibit or destroy microorganisms present on external or internal body parts.

Historically, Louis Pasteur (1822–1895) contributed to both areas. In his early research, he discovered that beer and wine making entailed a fermentation process involving initial growth of yeast in the fermentation of liquor. Later, he showed that a sterile broth infusion in a swan-necked flask showed no turbidity due to microbial growth (figure I.5.1), and that upon tilting the flask, the sterile infusion became readily contaminated. The swan-necked flask experiment was both classical and mon-

umental in that it helped resolve a debate, of more than 150 years, over the possible origin of microorganisms by **spontaneous generation** (abiogenesis).

The debate was finally squelched by John Tyndall, a physicist, who was able to establish an important fact overlooked by Pasteur—namely, that some bacteria in hay infusions existed in two forms: a **vegetative** form readily susceptible to death by boiling of the hay infusion, and a resting form now known as an **endospore,** which was resistant to boiling. With this knowledge, Tyndall developed a physical method of sterilization, which we now describe as **tyndallization,** whereby both vegetative cells and endospores are destroyed when the infusion is boiled intermittently with periods of cooling. For sterilization of some materials by tyndallization, temperatures below boiling are possible. (See Nester et al. *Microbiology: A human perspective,* Perspective 9.1.) Tyndallization, although a somewhat lengthy sterilization method, is sometimes used to sterilize chemical nutrients subject to decomposition by the higher temperatures of autoclaving.

At about this same time, chemical disinfectants for help in healing compound bone fractures were introduced by John Lister, an English surgeon, who was also impressed with Pasteur's findings. Lister had heard that carbolic acid (phenol) had remarkable effects when used to treat sewage in Carlisle; it not only prevented odors from farm lands irrigated with sewage, but also destroyed **entozoa,** intestinal parasites that usually infect cattle fed on such pastures.

The control of microbial growth has many applications today, both in microbiology and in such areas as plant and mammalian cell culture. Traditional examples of it include pure culture isolation, and preparation of sterile culture media, bandages, and instruments. It also includes commercial preparation of various microbial products such as antibiotics, fermented beverages, and food.

Exercises 7, 9, and 10 touch on maximization of microbial growth. In this section, the exercises deal with minimization or elimination of microbial growth by heat, ultraviolet light, osmotic pressure, antiseptics, and antibiotics. A final practical exercise relates to the use of preparative hand scrubbing for microbe elimination prior to active involvement in the operating room.

. . . we are too much accustomed to attribute to a single cause that which is the product of several, and the majority of our controversies come from that.

Von Liebig

---

pickles

EXERCISE 11

Moist and Dry Heat Sterilization: Thermal Death Point and Thermal Death Time

Getting Started

Physical methods used in the hospital environment to control microbial growth include heat and ultraviolet light to kill microorganisms; filtration to remove microorganisms from growth media containing heat-labile substances such as enzymes, and from air in operating rooms and cell transfer rooms; and the use of sterile gloves, masks, and clothing in these rooms to help control air convection of microorganisms present on skin and hair.

In this exercise, some effects of heat sterilization are studied since heat is commonly used to sterilize many of the materials used in hospitals and laboratories. When heat is applied, most microbes are killed, whereas when cold temperatures are applied, inhibition of microbial growth is more likely to occur. The sensitivity of a microorganism to heat is affected by its environment and genetics. Environment includes factors such as incubation temperature, chemical composition of the growth medium, and the age and concentration of cells in the growth medium. Genetically, some microorganisms are more tolerant of heat than others. Examples include the ability of microorganisms classified as **thermophiles** to grow at higher temperatures than others, and the ability of some microorganisms to produce heat-resistant structures called endospores. Members of the genera *Bacillus* and *Clostridium* are capable of endospore production.

Heat is applied in either a dry or moist form. Examples of dry heat include hot air ovens used largely for sterilization of glassware such as Petri dishes and pipets, and microincinerators used for sterilizing needles and loops. Dry heat kills by dehydrating microorganisms, which results in **irreversible denaturation** of essential enzyme systems. To sterilize with dry heat requires considerably more time and higher temperatures than with moist heat because dry heat penetrates to the inside of microbial cells more slowly than does moist heat. Typical sterilization times and temperatures are 2 hours at 165° C for dry heat and 15 minutes at 121° C for moist heat. The mode of action is the same for both.

Autoclaving is the most commonly used method of moist heat sterilization. Some other moist heat methods are **pasteurization, boiling,** and tyndallization. With autoclaving and tyndallization, both the vegetative and endospore forms of microorganisms are killed, whereas with pasteurization and boiling, usually only vegetative cells are killed.

Pasteurization, which is named for Louis Pasteur, is a moist heat process used in beverages such as milk, beer, and wine to kill **pathogenic** bacteria and reduce the number of nonpathogenic bacteria such as **thermoduric** bacteria. The beverages are heated under controlled conditions of temperature and time, either 63° C for 30 minutes or 72° C for 15 seconds. However, many endospore-forming bacteria survive pasteurization. The lower temperatures of pasteurization help preserve food flavor.

Tyndallization, named after the physicist John Tyndall, is sometimes used to sterilize nutrient media subject to inactivation by the higher temperatures of conventional autoclaving. It is also useful in emergencies, such as when an autoclave becomes inoperative. Tyndallization is a lengthy process requiring three days. The solution to be sterilized is usually steamed for 30 minutes in flowing steam (100° C) on each of three successive days. Between steaming times, the solution is left at room temperature. In principle, the first boiling kills all vegetative cells, the second boiling destroys newly germinated endospores, and the third boiling serves as an added insurance that no living cells remain in the solution.

Boiling for 10 minutes is used to rid solutions such as drinking water of vegetative forms of pathogenic bacteria and other pathogens such as parasitic worms and protozoa.

Two commonly used methods for determining the heat sensitivity of a microorganism are the **thermal death point** (TDP) and the **thermal death** time (TDT). The TDP is defined as the lowest temperature necessary to kill all of the microorganisms present in a culture in 10 minutes. The TDT is defined as the minimal time necessary to kill all of the microorganisms present in a culture held at a given temperature. These general principles are commonly used in establishing sterility requirements for many processes such as milk and food preservation, and for preparation of sterile media and hospital supplies.

Definitions

Autoclave.   A form of moist heat sterilization, conventionally performed at 121° C for 15 minutes.

Boiling.   Moist heat treatment that kills most pathogens. Conventionally, it is performed at 100° C; the time varies, although it is often done for 30 minutes.

Dry heat oven.   Dry heat sterilization, conventionally done at 160° C to 170° C for 2–3 hours.

Irreversible denaturation.   The destruction by heat of certain chemicals, including proteins such as enzymes.

Pasteurization.   The use of moist heat for beverages such as milk at a temperature sufficiently high enough to kill pathogens but not all organisms that may be present.

Pathogen.   An organism able to cause disease.

Thermal death point.   The lowest temperature able to kill all microbes in a culture after a given time.

Thermal death time.   The minimal time necessary to kill all microbes in a culture at a given temperature.

Thermoduric.   Microbes able to survive conventional pasteurization, usually 63° C for 30 minutes or 72° C for 15 seconds.

Thermophile.   An organism able to grow at temperatures above 55° C.

Tyndallization.   The process of using repeated cycles of heating and incubation to kill spore-forming bacteria.

Objectives

1. To provide background information about some physical sterilization methods requiring either moist heat or dry heat. Included in the discussion of moist heat methods are autoclaving, boiling, tyndallization, and pasteurization.
2. To introduce a quantitative laboratory method used for determining the susceptibility of different bacteria to the lethal effect of moist heat—thermal death point (TDP) and thermal death time (TDT).
3. To demonstrate the use of laboratory equipment commonly used for physical sterilization of moist and dry materials: the steam autoclave and the dry heat oven.

References

Nester et al. *Microbiology: A human perspective.* Chapter 9, 197–204.

Further Reading

Frazier, W. C., and Westhoff, D. *Food microbiology,* revised. New York: McGraw-Hill Book Company, 1988. Contains relevant information on principles of food preservation and spoilage.

Materials

Cultures
    24-hour 37° C *Staphylococcus epidermidis* culture in 5 ml TS broth, 4 tubes
    24-hour 37° C *Escherichia coli* culture in 5 ml TS broth, 4 tubes
    Spore suspension in 5 ml of sterile distilled water of a 4–5 day 37° C nutrient agar *Bacillus subtilis* culture, 4 tubes

5 ml of TS broth, 72 tubes

Two large beakers or cans, one for use as a water bath and the other for use as a reservoir of boiling water (per student)

Either a ring stand with wire screen and a Bunsen burner or a hot plate (per student)

One thermometer (per student)

A community water bath

Table 11.1 Culture and Heating Temperature Assignments (12 Students)

Bacterial Culture Assignment	Water Bath Temperature Assignment			
	40° C	60° C	80° C	100° C
Staphylococcus epidermidis	1	2	3	4
Escherichia coli	5	6	7	8
Bacillus subtilis	9	10	11	12

Procedure

First Session

Determination of Thermal Death Point and Thermal Death Time

Note: A procedural culture and heating distribution scheme for 12 students is shown in table 11.1. For this scheme, each student will receive one bacterial culture to be heated at one of the assigned temperatures (40° C, 60° C, 80° C, or 100° C).

1. Suspend your assigned culture by gently rolling the tube between your hands, followed by aseptically transferring a loopful to a fresh tube of broth (label "0 time Control").
2. Fill the beaker to be used as a water bath approximately half full with water, sufficient to totally immerse the broth culture without dampening the test-tube cap.
3. Place your tube of broth culture in the water bath along with an open tube of uninoculated broth in which a thermometer has been inserted for monitoring the water bath temperature.
4. Place the water bath and contents on either the ring stand or hot plate and heat to the assigned temperature. When the assigned temperature is reached, note the time, and remove the water bath from the heat source and place it on your bench top.
   (**Note:** Students with the 100° C assignment may wish to keep their water bath on the heat source, providing the water can be controlled at a low boil.)
   The temperature of the water bath can be maintained by periodically stirring in small amounts of boiling water obtained from the community water bath.

5. After 10 minutes of heating, resuspend the broth culture and aseptically transfer a loopful to a fresh tube of broth (label "10 minutes").
6. Repeat step 5 after 20, 30, and 40 minutes.
7. Write the initials of your culture as well your initials on all 5 tubes and incubate them in the 37° C incubator for 48 hours.

Demonstration of the Steam Autoclave and Hot Air Oven

The Steam-jacketed Autoclave

Note: As the instructor demonstrates the special features of the autoclave, follow the diagram in figure 11.1. Note the various control valves—their function and method of adjustment (exhaust valve, chamber valve, and safety valve); the steam pressure gauge; the thermometer and its location; and door to the chamber. Also make a note of special precautions necessary for proper sterilization.

1. Material sensitivity. Certain types of materials such as talcum powder, oil, or petroleum jelly cannot be steam sterilized because they are water repellent. Dry heat must be used instead. Some materials are destroyed or changed by the standard autoclave temperature of 121° C; for example, some sugars are hydrolyzed and some medications are chemically changed. In such instances, the autoclave may be operated at a lower pressure and temperature for a longer period of time. A heat-sensitive fluid material can normally be sterilized by filtration. Filtration removes the bacteria. Keep in mind that the smaller the filter pores, the slower the rate of filtration.
2. Proper preparation of materials. The steam must be in direct contact with all materials to be sterilized. Therefore, media container

Figure 11.1 Steam-jacketed autoclave. Entering steam displaces air downward and out through a port in the bottom of the chamber. Dry objects are placed in the autoclave in a position to avoid trapping air. Watery liquids generate their own steam.

closures such as metal caps with air passages, loosened screw cap lids, aluminum foil (heavy grade), and sometimes nonabsorbant cotton plugs are used.

3. Proper loading of supplies. There must be ample space between packs and containers so that the steam can circulate. When using cotton plugs, they should be loosely covered with foil to prevent moisture from the autoclave condensing on them during cooling.

4. Complete evacuation of air from the chamber. This is necessary before replacement of steam. Older models may require manual clearance, but in newer models it is automatic.

5. Proper temperature. Autoclaving at a pressure of 15 lb per square inch achieves a temperature of 121° C (250° F) at sea level. If the temperature gauge does not register this reading, trapping of colder air in the autoclave is indicated, which lowers the temperature.

Note: The temperature is critical, not the pressure.

6. Adequate sterilization time. After the chamber temperature reaches 121° C, additional time is required for the heat to penetrate the material. The larger the size of individual containers and packs, the more time required. Time must be adjusted to the individual load size. Also, highly contaminated materials require more time.

7. Completion of the autoclaving process. Rapid reduction in steam pressure can cause fluids to boil vigorously through the caps or to explode. Drop steam pressure gradually as cooling occurs and, when possible, allow material to dry in the autoclave. If removed while moist, the wrappings or plugs may provide a means for reentry of bacteria present in the room air.

The Hot Air Oven

The hot air, or dry heat, oven is used in most laboratories for both drying glassware and sterilization.

When using the dry heat oven, the following guidelines are important:

1. Types of material suitable for sterilization include oil, petroleum jelly, Vaseline, metal containers, and dry, clean glassware.
2. An oven with circulating air takes about half the sterilization time of a static air oven. Better heat transfer occurs with circulating air.
3. Proper packaging is necessary to assure air circulation to the inside surfaces. For example, syringes must be separate from the plunger so that all surfaces are exposed to circulating air.
4. Sterilization of dirty materials should be avoided. The presence of extraneous materials such as protein delays the process and may allow bacteria to survive inside the material.

5. Heat-sensitive tapes designed for the autoclave cannot be used to assure adequate sterilization because the hot air oven requires much higher temperatures than does the autoclave. Bacterial spores, however, can be used.
6. In the Laboratory Report, prepare a list of materials for your class that are sterilized in the autoclave. For each one, indicate the standard temperature, pressure, and time used for sterilization.

   Do the same for materials sterilized in the hot air oven. Indicate the oven sterilization temperature, time, and reasons for sterilizing there.

Second Session

After 48 hours, examine your broth tubes for the presence or absence of turbidity (growth). Write your results in the appropriate place in the table on the blackboard of your classroom. When all the results are entered, transfer them to table 11.2 of the Laboratory Report.

NOTES:

EXERCISE

11

Laboratory Report: Moist and Dry Heat Sterilization: Thermal Death Point and Thermal Death Time

Results

1. Determination of Thermal Death Point and Thermal Death Time

Table 11.2   Bacterial Growth at Assigned Temperatures and Times

Culture	40° C					60° C					80° C					100° C				
	C	10	20	30	40	C	10	20	30	40	C	10	20	30	40	C	10	20	30	40
S. epidermidis	ʃ	+	+	+	+	+	+	+	+	+		+	−	−	▽	−	−	−	−	−
E. coli		+	+	+	+		+	+	+	+		+	−	−	−		−	−	−	−
B. subtilis			+	+	+	+		+	+	+	+	+	+	+	+	+		−	−	−	−

Note: C = Control; 10, 20, 30, 40 = minutes; Use a + sign for growth and a − sign for no growth

    a. Determine the thermal death point for each culture:

Thermal death point (° C)

Staphylococcus epidermidis   ____

Escherichia coli   ____

Bacillus subtilis   ____

    b. Determine the thermal death time for each culture:

Thermal death time (minutes)

Staphylococcus epidermidis   ____

Escherichia coli   ____

Bacillus subtilis   ____

2. Evaluation of Materials Sterilized in Our Laboratory with Moist (Autoclave) and Dry Heat (Hot Air Oven)

   a. List of materials sterilized with the autoclave (see Procedure for criteria):

   b. List of materials sterilized with the hot air oven (see Procedure for criteria):

Questions

1. Was the *Bacillus subtilis* culture sterilized after 40 minutes of boiling? If not, what is necessary to assure sterility by boiling?

2. How would you set up an experiment to determine to the minute the TDT of S. *epidermidis*? Begin with the data you have already collected.

3. A practical question related to thermal death time (TDT) relates to a serious outbreak of E. *coli* infection in early 1993 in people who ate insufficiently grilled hamburgers. How would you set up an experiment to determine the TDT of a solid such as a hamburger? Assume that the thermal death point (TDP) is 67.2° C (157° F), the temperature required by many states to cook hamburger on an open grill. What factors would you consider in setting up such an experiment?

4. What is the most expedient method for sterilizing a heat-sensitive liquid that contains a spore-forming bacterium?

5. List one or more materials that are best sterilized by the following processes:

    a. Membrane filtration

    b. Ultraviolet light

    c. Dry heat

    d. Moist heat

    e. Tyndallization

    f. Radiation

6. What are three advantages of using metal caps rather than cotton for test-tube closures? Are there any disadvantages?

7. How would you sterilize a heat-sensitive growth medium containing thermoduric bacteria?

EXERCISE 12

Control of Microbial Growth with Ultraviolet Light

Figure 12.1 Germicidal activity of radiant energy.

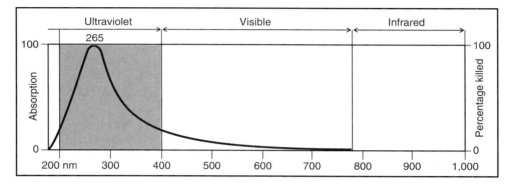

Getting Started

Ultraviolet (UV) light is the component of sunlight that is responsible for sunburn. It can also kill microorganisms by acting on their DNA and causing mutations. It consists of very short **nonionizing wavelengths of radiation** (200 to 400 nm) located just below blue light in the visible spectrum (figure 12.1).

The actual mechanism of mutation is the formation of pyrimidine dimers (figure 12.2). Two adjacent thymidines on a DNA molecule bind to each other; when the DNA is replicated, an incorrect base pair is frequently incorporated into the newly synthesized strand. This may cause mutation and if there is sufficient radiation, ultimately, the death of the cell.

UV does not penetrate surfaces and will not go through ordinary plastic or glass. It is only useful for killing organisms *on* surfaces and in the air. Sometimes UV lights are turned on in operating rooms and other places where airborne bacterial contamination is a problem. Since UV light quickly damages the eyes, these lights are turned on only when no one is in the irradiation area.

Bacteria vary in their sensitivity to UV. In this exercise, the sensitivity of *Bacillus* endospores will be compared with non-spore-forming cells. You will

Figure 12.2 Thymine dimer formation. Covalent bonds form between adjacent thymine molecules on the same strand of DNA. This distorts the shape of the DNA and prevents replication of the changed DNA. From Eugene W. Nester, C. Evans Roberts, and Martha T. Nester, *Microbiology: A Human Perspective.* Copyright © 1995 Wm. C. Brown Communications, Inc., Dubuque, Iowa. All Rights Reserved. Reprinted by permission.

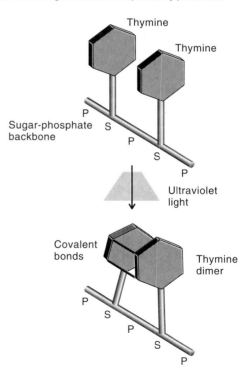

also irradiate a mixed culture such as organisms in saliva and hamburger to compare the resistance of different organisms. Immediately following irradiation it is important to place the samples in a dark container since some cells apparently killed by UV recover when exposed to longer wavelengths of light (light repair).

Definitions

Light repair.   DNA repair in cells previously exposed to UV by a DNA repair enzyme that requires visible light.

Nonionizing wavelengths of radiation.   Noncharged wavelengths of radiation with wavelengths between 175 nm and 350 nm. This light is invisible and is exemplified by UV light. Beta rays and X rays are examples of ionizing wavelengths of radiation.

Pyrimidine dimers.   Pyrimidine is a subunit of RNA and DNA. The three major pyrimidines are thymidine, cytosine, and uracil. A pyrimidine dimer is a molecule resulting from the covalent bonding of two identical adjacent pyrimidine molecules in the same nucleotide strand.

Thymidine.   One of the three major pyrimidines (see above).

Visible spectrum.   Uncharged wavelengths of radiation with wavelengths between 400 and 800 nm.

References

Nester et al. *Microbiology: A human perspective*. Chapter 9, 157–160, 198.

Materials

Cultures
Suspension of *Bacillus* spores in sterile saline
E. coli in TS broth
Raw hamburger mixed with sterile water

TS agar plates, 3 per team

UV lamp with shielding. An 18- to 36-inch fluorescent bulb is ideal. It enables uniform exposure of three to six partially opened Petri dishes. See Notes to the Instructor.

Sterile swabs, 3

Safety glasses for use with the UV lamp

Dark box for storing plates after UV exposure

Procedure

First Session

Safety Precautions: (1) The area for UV irradiation should be in an isolated part of the laboratory. (2) Students should wear safety glasses as a precautionary measure when working in this area. (3) Never look at the UV light after turning it on since it could result in severe eye damage. Skin damage is also a slight possibility.

1. Dip a sterile swab in a suspension of *Bacillus* spores and swab an agar plate in three directions as shown in figure 14.3.
2. Repeat the procedure with a suspended *E. coli* broth culture.
3. For the third plate, you can either moisten a swab with saliva or dip the swab into a mixture of sterile water and hamburger.

4. Place the plates under a UV lamp propped up about 8 to 10 inches from the bench surface. Open the Petri dish(es) and partially cover each plate with the lid. The part of the plate protected by the lid will be the control because UV does not penetrate most plastic.

5. Put on your safety glasses and turn on the UV light. Expose the plates to UV for 3 minutes.

6. Turn off the UV light. Cover the plates, invert them, and place them in a covered container. Incubate at 37° C for 48 hours.

Second Session

Observe the plates and record your findings in the Laboratory Report.

NOTES:

EXERCISE 12

Laboratory Report: Control of Microbial
Growth with Ultraviolet Light

Results

1. Record your observations for control and treated sides of Petri dishes exposed to UV light at a distance of _____ inches for _____ minutes:

    a. Plate containing *Bacillus* spores.

    most

    b. Plate containing *E. coli*.

    middle

    c. Plate containing either saliva or a raw hamburger suspension. Indicate which one you used.

    least

2. Which organisms were most resistant to UV? _____ Least resistant? _____

Questions

1. Why can't you use UV to sterilize microbiological media, e.g., agar or broth?

2. What is one way in which you can increase the sensitivity of UV radiation for use in sterilizing the laboratory bench surface in an inoculation room?

3. Rate these particles in order of sensitivity to UV radiation, where 1 is the most sensitive and 4 the least sensitive:

___ *Bacillus* endospore

___ *Bacillus* vegetative cell

___ Naked virus DNA

___ *Bacillus* protoplast

4. Frequently organisms isolated from the environment are pigmented, while organisms isolated from the intestine or other protected places are not. Can you provide an explanation for this?

NOTES:

Ionizing radiation — passes through substances

Gamma — nucleus

X — rays — e—

ultraviolet : 210 - 300 nm   nucleic acids — does not penetrate
                                                        surfaces

EXERCISE 13

Osmotic Pressure, and the Rate and Amount of Microbial Growth

Getting Started

Osmosis, which is derived from the Greek word "alter," refers to the process of flow or diffusion that takes place through a **semipermeable membrane.** (In a living cell, the cytoplasmic membrane located adjacent to the inside of the cell wall represents such a membrane.)

Both the cytoplasmic membrane and the cell wall help prevent the cell from either bursting **(plasmoptysis)** or collapsing **(plasmolysis)** due to either entry or removal of water from the cell, respectively. The **solute** concentration outside the cell determines which, if any, process happens. When the solute concentration inside the cell is the same as the solute concentration on the outside of the cell **(isotonic),** the cell remains intact. When the solute concentration outside the cell is less than the solute concentration inside the cell, an inward **osmotic pressure** occurs, and water enters the cell in an attempt to equalize the solute concentration on either side of the cytoplasmic membrane. If the solute concentration outside the cell is sufficiently low **(hypotonic),** the cell will absorb water and sometimes burst (plasmoptysis). However, it rarely occurs due to the rigidity of the cell wall. The reverse phenomenon, cell shrinkage followed by cell lysis (plasmolysis), can occur when the cell is placed in a more concentrated **(hypertonic)** solution. This can become a life-and-death problem if too much water is removed from the cells (figure 13.1).

When placed in an isotonic solution, some cells recover, although there are many genera that die once the external osmotic pressure exceeds their limitations. This concept is the basis of food preservation—the use of high salt concentrations (for cheese and pickle brine) and sugar concentrations (in honey and jams).

In general, fungi (yeasts and molds) are much more resistant to high external solute concentrations than are bacteria, which is one reason fungi

Figure 13.1 Movement of water into and out of cells. (*a*) Low and (*b*) very high salt-containing solutions. The cytoplasmic membrane is semipermeable and only allows water molecules to pass through freely. From Eugene W. Nester, C. Evans Roberts, and Martha T. Nester, *Microbiology: A Human Perspective.* Copyright © 1995 Wm. C. Brown Communications, Inc., Dubuque, Iowa. All Rights Reserved. Reprinted by permission.

(a) Water molecules move into the cell to equalize the pressure and push the cytoplasmic membrane against the cell wall (plasmoptysis).

(b) Water molecules leave the cell and the cytoplasmic membrane shrinks from the cell wall (plasmolysis).

can grow in or on jelly, cheese, and fruit. There are, of course, exceptions among the bacteria; for example, the genus *Halobacterium* in the *Archaebacteria* and the genus *Micrococcus halobius* in the family *Micrococcaceae,* which is sometimes found in nature growing on highly salted (25–32%) protein products such as fish and animal hides. It is interesting to note that all of these bacteria produce a red pigment.

There are also some salt-tolerant *Staphylococcus* strains able to grow at salt concentrations greater than 10% (w/v), allowing them to grow on skin surfaces. The salt-loving (**halophilic**) bacteria are also unique in that, like all *Archaebacteria*, they lack muramic acid as a bonding agent in their cell walls. Instead, their cell walls are believed to contain sodium and potassium ions. These ions help confer cell wall rigidity, perhaps helping explain the reason they require such high salt concentrations for good growth.

Yeasts and molds able to grow in high sugar (50–75%) and sometimes high salt concentrations (25–30%) are termed **saccharophilic** and halophilic fungi, respectively. Some of the yeast and fungus genera included are *Debaromyces*, *Saccharomyces*, *Zygosaccharomyces*, *Aspergillus*, and *Penicillium*.

In this exercise, you determine the ability of some of the previously mentioned halophiles and saccharophiles to grow on the surface of TYEG agar containing increasing concentrations of salt and sugar. *Escherichia coli* is added as a salt-sensitive gram-negative rod control. You also have an opportunity to examine any changes in cell structure with increasing salt and sugar concentrations.

It should be kept in mind that all of these halophilic and saccharophilic microbes are characterized by an increase in lag time and a decrease in growth rate, and in the amount of cell substance synthesized. In some ways, their **growth curve** (see exercise 10) parallels what happens when they are grown at a temperature below their optimal growth temperature. For example, halobacteria have a **generation time** of 7 hours and halococci of 15 hours.

Definitions

Generation time.   The time required for one cell to divide into two cells.

Growth curve.   A curve describing the four readily distinguishable phases of microbial growth: lag, log, stationary, and death.

Halophilic microbes.   A salt-requiring organism able to grow in a medium containing a salt concentration high enough to inhibit other organisms.

Hypertonic.   A fluid having an osmotic pressure greater than another fluid with which it is compared.

Hypotonic.   A fluid having an osmotic pressure lower than another fluid with which it is compared.

Osmotic pressure.   The pressure exerted by water on a membrane as a result of a difference in the concentration of solute molecules on each side of the membrane.

Plasmolysis.   Contraction or shrinking of the cytoplasmic membrane away from the cell wall due to a loss of water from the cell.

Plasmoptysis.   The bursting of protoplasm from a cell due to rupture of the cell wall by absorbing excess water from the external environment.

Saccharophilic microbes.   Microbes able to grow in environments containing high sugar concentrations.

Semipermeable membrane.   A membrane permitting the passage of some materials but not others through a membrane (for example, the cytoplasmic membrane). Usually depends on the size of the molecule.

Objectives

1. To provide an introduction to osmotic pressure and show how it may be used to inhibit growth of less osmotolerant microbes, while allowing more osmotolerant microbes to grow, although often at a considerably slower growth rate.
2. To show that some microorganisms either require or grow better in an environment containing high concentrations of salt (halophilic) or sugar (saccharophilic).

References

Mossel, D. A. A. "Ecological Essentials of Antimicrobial Food Preservation," pp. 177–195 in *Microbes and biological productivity*. Edited by D. E. Hughes and A. H. Rose. Cambridge University Press, 1971.

Nester et al. *Microbiology: A human perspective*. Chapter 3 (Functional Anatomy of Prokaryotes and Eukaryotes), chapter 9 (Control of Microbial Growth), and chapter 33 (The Microbiology of Food and Beverages).

Materials

Cultures (may be shared by 2 to 4 students)

Use TYEG salts agar slants for *Escherichia coli*, *Micrococcus luteus*, and *Saccharomyces cerevisiae*.

Use American Type Culture Collection (ATCC) medium 213 for *Halobacterium salinarium*.

Incubate *E. coli*, *M. luteus*, and *S. cerevisiae* cultures for 24 hours at 35° C. Incubate *H. salinarium* culture for 1 week (perhaps longer) at 35° C.

TYEG salts agar plates containing 0.5, 5, 10, and 20% NaCl, 4 plates.

TYEG salts agar plates containing 0, 10, 25, and 50% sucrose, 4 plates.

Procedure

First Session

Note: One student of the pair can inoculate the 4 TSA plates containing 0.5, 5, 10, and 20% NaCl, while the other student inoculates the 4 plates containing 0, 10, 25, and 50% sucrose.

1. First, use a glass-marking pencil to divide the undersurface of the eight plates in quadrants and label with the initials of the four test organisms; for example, *E.c.* for *Escherichia coli*, etc. Also label the underside of each plate with the salt or sugar concentrations and your name.

2. Using aseptic technique, remove a loopful from a culture and streak the appropriate quadrant of your plate in a straight line approximately 1 inch long. Then reflame your loop, cool it for a few seconds, and make a series of cross streaks approximately one-half inch long in order to initiate single colonies for use in studying colonial morphology (figure 13.2). Repeat inoculation procedure for culture #1 in the appropriate quadrant of the remaining three agar plates.

Figure 13.2 Streaking procedure for isolation of single colonies.

Original streak

Cross streaks

3. Repeat the inoculation procedure for the remaining three test organisms.
4. Invert and incubate the eight plates at 30° C.
5. Observe the plates periodically (up to 1 week or more if necessary) for growth.

Second Session

1. Examine your plates for the presence (+) or absence (−) of growth. For growth, use 1 to 3 + signs (+=minimal, ++=some, and +++=good). Enter results in table 13.1 (various salt concentrations) and table 13.2 (various sugar concentrations) of the Laboratory Report.

2. Compare the colonial growth characteristics of control cultures with those grown on increasing salt and sugar concentrations. Make notes of any marked changes in colony color, colony size (in mm), and colony texture: dull or glistening, rough or smooth, and flat or raised. Record your findings in the Laboratory Report.

3. Prepare wet mounts of bacteria and yeast colonies showing marked changes in visual appearance from the controls. Examine bacteria with the oil-immersion objective and yeasts with the high dry objective. Look for plasmolyzed cells and other changes such as in cell form and size. Prepare drawings of any such changes in the Laboratory Report.

NOTES:

Osmotic Pressure – Osmosis

  Hypotonic – greater H_2O – lose (solution)

  Hypertonic – lower H_2O – gains (solution)

  semipermeable membrane

  equal – isotonic

  plasmolysis – shriveling / dehydration of cells – hypertonic

  plasmoptysis – expanding and bursting of cells – hypotonic

    bacteria like greater concentration on outside

    hypertonic – no growth

    halophiles – high salt concentrations

    yeasts + molds – sugar

    TVEG agar –   one of each plate

    parafilm plates

EXERCISE

Laboratory Report: Osmotic Pressure, and the Rate and Amount of Microbial Growth

Results

1. Examination of Petri dish cultures for the presence (+) or absence (−) of growth in the presence of increasing salt (table 13.1) and sugar concentrations (table 13.2).

Table 13.1   Presence or Absence of Growth on TYEG Agar Plates Containing NaCl and Incubated for 48 hours to 1 week

Culture	NaCl Concentration (%)							
	0.5		5		10		20	
	48hr	1wk	48hr	1wk	48hr	1wk	48hr	1wk
Escherichia coli	+		+		+		−	
Halobacterium salinarium	−		−		−		−	
Micrococcus luteus	+		+				+	
Saccharomyces cerevisiae	+		+		+		+	

Table 13.2   Presence or Absence of Growth on TYEG Agar Plates Containing Sucrose and Incubated for 48 hours to 1 week

Culture	Sucrose Concentration (%)							
	0		10		25		50	
	48hr	1wk	48hr	1wk	48hr	1wk	48hr	1wk
Escherichia coli	+		+		+		+	
Halobacterium salinarium	−		−				−	
Micrococcus luteus	+		+		+		+	
Saccharomyces cerevisiae	+		+		+		+	

2. Comparison of the colonial growth characteristics of control cultures with those grown on increasing salt and sugar concentrations:

   a. *Escherichia coli*

NaCl %	Colony color	Colony size	Colony texture
0.5			
5			
10			
20			

Sucrose %

0

10

25

50

   b. *Halobacterium salinarium*

NaCl %	Colony color	Colony size	Colony texture
0.5			
5			
10			
20			

Sucrose %

0

10

25

50

   c. *Micrococcus luteus*

NaCl %	Colony color	Colony size	Colony texture
0.5			
5			
10			
20			

Sucrose %

0

10

25

50

      d. *Saccharomyces cerevisiae*

NaCl %	Colony color	Colony size	Colony texture
0.5			
5			
10			
20			

Sucrose %

0

10

25

50

3. Wet mounts of bacteria and yeast colonies showing *marked* changes in visual appearance from the salt and sugar agar media controls:

    a. *Escherichia coli*
    b. *Halobacterium salinarium*
    c. *Micrococcus luteus*
    d. *Saccharomyces cerevisiae*

Questions

1. From your studies, which organism(s) tolerate salt best? _____ Least? _____

2. What organism(s) tolerate sugar best? _____ Least? _____

3. Compare bacteria and yeast with respect to salt tolerance. Bear in mind both colonial and cellular appearance in formulating your answers.

4. Compare bacteria and yeast with respect to sugar tolerance. Bear in mind both colonial and cellular appearance in formulating your answer.

5. What evidence did you find of a *nutritional requirement* for salt or sugar in the growth medium?

6. Matching
   Each answer may be used one or more times.

   1. *Halobacterium* ____ osmosensitive

   2. *Saccharomyces* ____ long generation time

   3. *Escherichia coli* ____ saccharophilic

   4. *Micrococcus* ____ osmotolerant

7. Matching
   Choose the best answer. Each answer may be used one or more times, or not at all.

   1. Plasmolysis ____ isotonic solution

   2. Plasmoptysis ____ hypotonic solution

   3. Normal cell growth ____ hypertonic solution

   ____ swelling of cells

EXERCISE 14

Antiseptics and Antibiotics

Getting Started

It was during a visit through Central Europe in 1908 that I came across the fact that almost every farmhouse followed the practice of keeping a moldly loaf on one of the beams in the kitchen. When asked the reason for this I was told that this was an old custom and that when any member of the family received an injury such as a cut or bruise a thin slice from the outside of the loaf was cut off, mixed into a paste with water and applied to the wound with a bandage. I was assured that no infection would then result from such a cut.

Dr. A. E. Cliffe, a Montreal biochemist

Historically, chemicals such as antibiotics have been in existence a long time. However, the therapeutic properties of antibiotics simply were not recognized as such until Alexander Fleming's discoveries in the 1930s.

By definition, **antibiotics** are chemicals produced and secreted by microorganisms (bacteria, fungi, and actinomycetes) that can inhibit or destroy the growth of **pathogenic** microorganisms, often by altering an **essential metabolic pathway.** To be effective in human medicine, ideally they should be nontoxic to the human host and should discourage the formation of microbial strains resistant to the antibiotic (see exercise 15). Many antibiotics in use today can also be synthesized in the chemical laboratory.

Another interesting group of synthetic chemicals that act as antimetabolites are sulfa drugs, which originate from the azo group of dyes. One of these drugs, marketed under the name of Prontosil, inhibited microbial growth when tested *in vivo* (initially in mice and later in man), but when tested *in vitro* (test tubes) against streptococci nothing happened. In the 1930s, scientists at the Pasteur Institute in Paris showed that if Prontosil was chemically reduced it was not only active in the body, but also able to kill

Figure 14.1 (*a*) Structures of sulfanilamide (sulfa drug) and of para-aminobenzoic acid (PABA). The portions of the molecules that differ from each other are shaded. (*b*) Reversible competitive inhibition of folic acid synthesis by sulfa drug. The higher the concentration of sulfa drug molecules relative to PABA, the more likely that the enzyme will bind to the sulfa drug, and the greater the inhibition of folic acid synthesis. From Eugene W. Nester, C. Evans Roberts, and Martha T. Nester, *Microbiology: A Human Perspective.* Copyright © 1995 Wm. C. Brown Communications, Inc., Dubuque, Iowa. All Rights Reserved. Reprinted by permission.

Table 14.1   Chemical Compounds Commonly Used in Hospitals for Controlling Growth of Microorganisms

Sodium hypochlorite (5%)	Disinfectant	External surfaces, such as tables
Iodine (1% in 70% alcohol)	Disinfectant	External surfaces, such as tables
Iodophors (70 ppm avail. I_2)	Disinfectant	External surfaces, such as tables
Lysol (5%), a solution of cresol with soap	Disinfectant	External surfaces, such as tables
Phenol (5%), carbolic acid, source coal tar	Disinfectant	External surfaces, such as tables
Hexachlorophene (pHisoHex, Dial soap)	Disinfectant	Presurgical hand washing
Formaldehyde (4%)	Disinfectant	Oral and rectal thermometers
Iodophors (70 ppm avail. I_2)	Disinfectant	Oral and rectal thermometers
Zephrin (.001%)	Disinfectant	Oral and rectal thermometers
Alcohol, ethanol (70%)	Antiseptic	Skin
Iodine (tincture in alcohol with KI)	Antiseptic	Skin
Iodophors	Antiseptic	Skin
Organic mercury compounds (merthiolate, mercurochrome)	Antiseptic	Skin
Hydrogen peroxide (3%)	Antiseptic	Superficial skin infections
Potassium permanganate	Antiseptic	Urethral, superficial skin fungus infections
Silver nitrate (1%) (Argyrol)	Antiseptic	Prevention of eye infections in newborn babies
Zinc oxide paste	Antiseptic	Diaper rash
Zinc salts of fatty acids (Desenex)	Antiseptic	Treatment of athlete's foot
Glycerol (50%)	Antiseptic	Prevent bacterial growth in stool and surgical specimens
Ethylene oxide gas (12%)	Sterilization	Linens, syringes, etc.
Formaldehyde (20% in 70% alcohol)	Sterilization	Metal instruments
Glutaraldehyde (pH 7.5 or more)	Sterilization	Metal instruments

bacteria *in vitro*. Prontosil was active *in vivo*, (but not *in vitro*) because enzymes within mice and men reduce Prontosil to a smaller active molecule, known as sulfanilamide (figure 14.1). The inhibitory action of the sulfanilamide-type drugs is one of **competitive inhibition,** in which the sulfanilamide acts as an antimetabolite. The sulfanilamide component replaces para-aminobenzoic acid (PABA), an **essential metabolite,** as a necessary part of folic acid, an essential **coenzyme** involved in amino acid synthesis.

Antiseptics are chemicals able to inhibit *in vivo* **sepsis** (infection). They do not need to kill the sepsis-producing agent, merely inhibit its growth. Antiseptic chemicals must be sufficiently nontoxic to allow application to skin and mucous membranes, such as the use of Listerine for gargling.

These same chemicals act as **disinfectants** (a chemical able to kill vegetative forms but not necessarily spore forms of bacteria) when used at higher concentration levels. Toxicity is a major factor in determining usage of a chemical as either an antiseptic or disinfectant. Most *inorganic* heavy metal compounds can be used only as disinfectants (for example, mercurials). *Organic* heavy metal mercurial compounds (like mercurochrome) and also quaternary ammonium compounds and phenolics can be used as either antiseptics or disinfectants, depending on the concentration used.

The mode of action of antiseptics and disinfectants differs from antibiotics in that they act largely by denaturing proteins. They also lower surface tension, which is accompanied by cytoplasmic membrane dissolution (quaternary ammonium compounds), and act as oxidizing agents (chlorine-containing chemicals). A list of commonly used antiseptics and disinfectants and their area of application is shown in table 14.1.

An assay still used in many clinical laboratories to test the potency of antibiotics and drugs is a filter paper disc-agar diffusion procedure commonly known as the Kirby-Bauer test. A filter paper disc-agar diffusion method is also used for

Table 14.2 Chart Containing Zone Diameter Interpretive Standards for Determining the Sensitivity of Bacteria to Antimicrobial Agents*

Antimicrobial Agent	Disc Content	Zone Diameter, Nearest Whole mm		
		Resistant	Intermediate	Susceptible
Penicillin				
when testing staphylococci[b,c]	10 units	≤28	–	≥29
when testing enterococci[d]	10 units	≤14	–	≥15[d]
when testing streptococci[e]	10 units	≤19	20–27	≥28
Streptomycin				
when testing enterococci for high level resistance	**300 µg**	**6**	**7–9[q]**	**≥10**
when testing other organisms	**10 µg**	**≤11**	**12–14**	**≥15**
Tetracycline[m]	30 µg	≤14	15–18	≥19
Chloramphenicol	30 µg	≤12	13–17	≥18
Sulfonamides[n,o]	300 µg	≤12	13–16	≥17
Erythromycin	15 µg	≤13	14–22	≥23

National Committee for Clinical Laboratory Standards. *Performance Standards for Antimicrobial Disk Susceptibility Tests,* Fifth Edition; Approved Standard. Copyright © 1993 The National Committee for Clinical Laboratory Standards. By permission.
NOTE: Information in boldface type is considered tentative for one year.
[b]Resistant strains of *Staphylococcus aureus* produce β-lactamase and the testing of the 10-unit penicillin disc is preferred. Penicillin should be used to test the susceptibility of all penicillinase-sensitive penicillins, such as ampicillin, amoxicillin, azlocillin, bacampicillin, hetacillin, carbenicillin, mezlocillin, piperacillin, and ticarcillin. Results may also be applied to phenoxymethyl penicillin or phenethicillin.
[c]Staphylococci exhibiting resistance to methicillin, oxacillin, or nafcillin should be reported as also resistant to other penicillins, cephalosporins, carbacephems, carbapenems, and β-lactamase inhibitor combinations despite apparent *in vitro* susceptibility of some strains to the latter agents. This is because infections with methicillin-resistant staphylococci have not responded favorably to therapy with β-lactam antibiotics.
[d] The "Susceptible" category for penicillin or ampicillin implies the need for high-dose therapy for serious enterococcal infections. If possible, this should be denoted by a footnote on the susceptibility report form. Enterococcal endocarditis requires combined therapy with high-dose penicillin or high-dose ampicillin, or vancomycin, or teicoplanin plus gentamicin or streptomycin for bactericidal action. Since ampicillin or penicillin resistance among enterococci due to β-lactamase production is not reliably detected using routine disc or dilution methods, a direct, nitrocefin-based β-lactamase test is recommended. Synergy between ampicillin, penicillin, or vancomycin and an aminoglycoside can be predicted for enterococci by using a high-level aminoglycoside screening test.
[e]A penicillin MIC should be determined on isolates of *viridans Streptococcus* from patients with infective endocarditis.
[m]Tetracycline is the class disc for all tetracylines, and the results can be applied to chlortetracycline, demeclocycline, doxycycline, methacycline, minocycline, and oxytetracycline. However, certain organisms may be more susceptible to doxycycline and minocycline than to tetracycline (such as some staphylococci and *Acinetobacter*).
[n]Susceptibility data for cinoxacin, nalidixic acid, nitrofurantoin, norfloxacin, sulfonamides, and trimethoprim apply only to organisms isolated from urinary tract infections.
[o]The sulfisoxazole disc can be used to represent any of the currently available sulfonamide preparations. Blood-containing media except for lysed horse blood are generally not suitable for testing sulfonamides or trimethoprim. Mueller-Hinton agar should be checked for excessive levels of thymidine as described in table 3.
[q]If the zone is 7 to 9 mm, the test is inconclusive and an agar dilution or broth microdilution screen test should be performed to confirm resistance.

determining the potency of antiseptics. In this exercise, you will have an opportunity to determine both antiseptic and antibiotic potency with a modified Kirby-Bauer test.

Discs of filter paper impregnated with antibiotic solutions in the same range of concentrations obtainable in the human body are placed on an agar plate heavily seeded with the test bacterium. When incubated, the bacteria grow in a smooth lawn of confluent growth except in a clear zone around the antibiotic disc. The clear growth inhibition zone does not necessarily indicate degree of microbial susceptibility to the antibiotic. Zone size itself does not indicate if the antibiotic is appropriate for use in clinical treatment. When conducting the precisely

controlled Kirby-Bauer test, various special conditions such as 2–5 hour cultures, controlled inoculum size, and short incubation periods are necessary. Such conditions are difficult to achieve in the time frame of most classrooms. Consequently, the simpler test performed in this exercise will demonstrate the Kirby-Bauer test principle without the added restrictions. Upon observation, the diameter of the clear zone of inhibition surrounding each antibiotic disc can be compared to that contained for the antibiotic on a standard chart (table 14.2). From this chart, one can determine if the test bacterium is **resistant, intermediate,** or **sensitive** to the antibiotic in question. A comparison chart is not available for antiseptics.

Definitions

Antibiotic. A chemical produced largely by certain bacteria and fungi that can inhibit or destroy the growth of other organisms including pathogenic microorganisms.

Antimetabolite. A substance that inhibits the utilization of a metabolite necessary for growth (see figure 14.1).

Antiseptic. A chemical that inhibits or kills microbes. The definition also implies that the chemical is sufficiently nontoxic that it may be applied to skin and mucous membranes.

Coenzyme. Any heat-stable, nonprotein compound that forms an active portion of an enzyme system after combination with an enzyme precursor (apoenzyme). Many of the B vitamins are coenzymes.

Competitive inhibition. The inhibition of enzyme activity caused by the competition between the inhibitor and the substrate for the active (catalytic) site on the enzyme.

Disinfectant. A chemical agent that rids an area of pathogenic microorganisms. In so doing, it kills vegetative forms of bacteria but ordinarily not spore forms. The definition also implies that the chemical is sufficiently toxic that it should not be applied to body surfaces, only to material objects.

Essential metabolic pathway. A pathway necessary for growth; if inhibited, the organism usually dies.

Pathogen. Any agent capable of causing disease, usually a microorganism.

Objectives

1. To provide introductory information about the origin and usage of antibiotics, antiseptics, and disinfectants.
2. To provide you with an opportunity to evaluate the bacteriostatic activity of antibiotics, antiseptics, and disinfectants with a modified Kirby-Bauer test.

References

Nester et al. *Microbiology: A human perspective.* Chapter 9, 198, 205–212 (antiseptics and disinfectants) and chapter 29, Perspective 29.4, 645–648, 655–660 (antibiotics).

Materials

Cultures (per team of 2 to 4 students)
Bacteria (24 hour 37° C TS broth cultures)
  Staphylococcus epidermidis (a gram-positive coccus)
  Escherichia coli (a gram-negative rod)
  Pseudomonas aeruginosa (nonfermenting gram-negative rod)
  Mycobacterium smegmatis (acid-fast rod)

Vials of the following antibiotic discs, or dispensers:

Penicillin, 10 μg; streptomycin, 10 μg; tetracycline, 30 μg; chloramphenicol, 30 μg; nystatin, 100 units; sulfanilamide (or another sulfonamide), 300 μg; erythromycin, 15 μg.

Beakers containing 10 ml aliquots of the following chemicals: 70% ethanol, 3% hydrogen peroxide, antiseptic mouthwash such as Listerine, and 0.3% hexachlorophene.

Mueller-Hinton agar, 6 plates
Sterile cotton swabs, 6
Sterile filter paper discs, 1/4″ or 1/2″, 8
Small forceps, 1 per student
Ruler divided in mm

Procedure

Filter Paper Disc Technique for Antiseptics and Disinfectants: First Session

1. With a glass-marking pencil, divide the underside of two plates of Mueller-Hinton agar into quadrants and label them 1 through 4.

2. Label the cover of one Petri dish *S. epidermidis* and the cover of the other dish *E. coli*.

3. Suspend the *S. epidermidis* culture, then insert and moisten a sterile swab, remove excess, followed by streaking the swab in all directions on the surface of the agar plate. Discard swab in the appropriate waste container.

4. Repeat step 3 with *E. coli*.

5. Sterilize forceps by dipping them in 95% alcohol and then touch to the flame of the Bunsen burner. Air cool.

6. Using forceps, remove one of the filter paper discs from the container and dip it into solution 1: 70% ethanol.

7. Drain the disc thoroughly on a piece of clean absorbent toweling and place it in the center of quadrant 1 of the dish labeled *S. epidermidis* (figure 14.2). Tap disc gently.

8. Repeat steps 4, 5, and 6 and place the disc in the center of quadrant 1 of the plate labeled *E. coli*.

9. Repeat steps 4 through 7 for the remaining three compounds, using first 3% hydrogen peroxide, then the antiseptic mouthwash, and last hexachlorophene.

10. Invert the Petri dishes and incubate at 37° C for 48 hours.

Filter Paper Disc Technique for Antibiotics: First Session

1. Divide the four broth cultures among team members, so that each student sets up at least one susceptibility test.

2. With a glass-marking pencil, divide the underside of four plates of Mueller-Hinton agar into six pie-shaped sections (figure 14.3b).

3. Label the cover of each plate with the name of the respective bacterium (see Materials for names).

4. Using aseptic technique (see section A step 3 above), streak the first broth culture as shown in figure 14.3a. The remaining three cultures should be streaked on separate plates in a similar manner.

5. Assemble the seven antibiotic discs and record the code of each in table 14.4 of the Laboratory Report so that the results can be interpreted easily in Session Two.

Figure 14.2 (*a–d*) Filter paper disc technique for antiseptics.

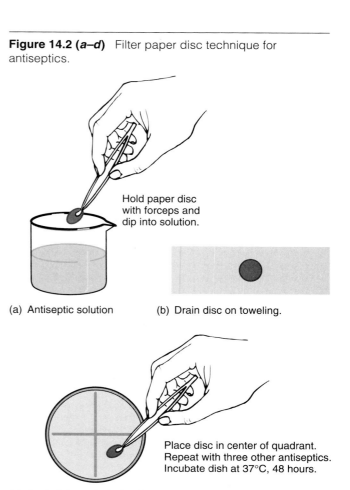

(a) Antiseptic solution (b) Drain disc on toweling.

(c) Petri dish seeded with *S. aureus* or *E. coli*

Hold paper disc with forceps and dip into solution.

Place disc in center of quadrant. Repeat with three other antiseptics. Incubate dish at 37°C, 48 hours.

(d) Measure the zone of inhibition surrounding each disc.

6. Heat sterilize forceps (see step 5 above), and remove an antibiotic disc from the container. Place gently, with identification side up, in the center of one of the pie-shaped sections of the agar plate (see figure 14.3b). Tap gently to fix in position.

7. Continue placing the remaining six antibiotic discs in the same way, putting the last disc in the center where the lines cross.

Figure 14.3 Antibiotic susceptibility test. (*a*) Procedure for streaking an agar plate in three or more directions with a swab inoculum in order to achieve a uniform lawn of growth. (*b*) The underside of a Mueller-Hinton agar plate showing the marking of sections and the arrangement for placement of antibiotic discs on the agar surface.

(a)

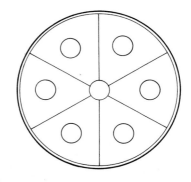

(b)

Note: Be sure to flame the forceps with alcohol after placing each disc because it is possible to contaminate stock vials with resistant organisms.

Note: If a disc dispenser is used, follow the manufacturer's instructions.

8. Repeat steps 4 through 6 with the remaining three cultures.
9. Invert and incubate the plates at 37° C for 48 hours.

Filter Paper Disc Technique for Antiseptics and Disinfectants: Second Session

1. Turn over the *S. epidermidis* plate and with a ruler calibrated in mm, determine the diameter of the clear zone surrounding each disc. Repeat with the *E. coli* plate.

   Note: It may be necessary to illuminate the plate in order to define the clear zone boundary.

2. Record your results in table 14.3 of the Laboratory Report.

Filter Paper Disc Technique for Antibiotics: Second Session

1. Observe plates using the same method described in step 1 above for antiseptics and disinfectants. In addition, make note of any large colonies present in the clear zone of growth inhibition surrounding each antibiotic disc. They may be resistant mutants.
2. Record your findings in table 14.4 of the Laboratory Report.
3. Compare your results where possible with table 14.2 and indicate in table 14.4 the susceptibility of your test cultures (*when possible*) to the antibiotics as resistant (R), intermediate (I), or sensitive (S).

Note: Your answers may not agree exactly with those in table 14.2 since this is a modified Kirby-Bauer test.

EXERCISE

Laboratory Report: Antiseptics and Antibiotics

Results

1. Filter Paper Disc Technique for Antiseptics and Disinfectants

Table 14.3   Bacteriostatic Activity of Various Antiseptics and Disinfectants

Antiseptic	Zone of Inhibition (mm)	
	Staphylococcus epidermis	*Escherichia coli*
70% ethanol		
3% hydrogen peroxide		
Listerine®		
0.3% hexachlorophene		
Others:		

What general conclusions can you make from this study? What differences, if any, did you observe on your plates between antiseptic and disinfectant preparations?

2. Filter Paper Disc Technique for Antibiotics

Table 14.4   Antibiotic Susceptibility (Modified Kirby-Bauer Test)

Test Organism	Zone of Inhibition (mm)							Susceptibility[a]	Notes
	[b]Chl	Ery	Pen	Str	Sul	Tet	Nys	(R, I, or S)	
S. epidermidis									
E. coli									
P. aeruginosa									
M. smegmatis									

[a]R = Resistant, I = Intermediate, and S = Susceptible.
[b]Chl = Chloramphenicol (Chloromycetin), 30 μg; Ery = Erythromycin, 15 μg; Pen = Penicillin G, 10 μg; Str = Streptomycin, 10 μg; Sul = Sulfanilamide, 300 μg; Tet = Tetracycline, 30 μg; Nys = Nystatin, 100 units.

Questions

1. What correlation did you find, if any, between the Gram-staining reaction of a microorganism and its susceptibility to antiseptics and disinfectants?

2. You may have noted that nystatin was not listed in table 14.2. The reason for its omission is that it is an antifungal antibiotic. Was it antibiotically active against any of the bacteria you studied? Is there an organism that could have been tested that might have been susceptible?

3. Name some other factors affecting the size of the zone of growth inhibition that were not included in your modified Kirby-Bauer test. Why were they omitted? Discuss their importance. The *Manual of Clinical Microbiology*, 5th edition, eds. Balows, Hauser, Herrman, Isenberg, and Shadomy, published by the American Society for Microbiology, Washington, D.C., 1991, is an excellent reference.

4. To what general groups of organic compounds does hexachlorophene belong? What are the advantages and disadvantages of using hexachlorophene for surgical scrub downs?

5. Matching

    1. 70% ethanol   ____ antibiotic

    2. 5% phenol   ____ antiseptic

    3. nystatin   ____ coal tar dye

    4. prontosil   ____ disinfectant

    5. sulfanilamide   ____ drug

Antibiotics- bacteria fungi

Synthetic drugs - sulfa → competitive inhibition (PABA)

cell wall injury → lysis, inhibit   cell membrane   cytoplasmic - denature proteins
cause DNA + RNA malfunction   (enzymes)

antiseptic - on living tissue

disinfectants - non-living

Kirby - Bauer : technique for streaking the plate

Mueller - Hinter : type of broth

② dip

Kirby - Bauer
technique

④ dip

sterilize forceps → filter paper → place in antiseptic → place on plate

clear area - zone of inhibition

E. Coli   mutations
S. Epidermitis   lethal
P. aeragirosa   plasmids - carry drug resistance
M. phlei

β - lactimase

INTRODUCTION to Microbial Genetics

In this section, three aspects of microbial genetics will be studied: selection of mutants, gene transfer, and gene regulation.

Selection of Mutants Mutations are constantly occurring in all living things. The replication of DNA is amazingly error-free but about once in every 100 million duplications of a gene a mistake is made, and the gene does not function in the normal way or produce the proper protein. Usually, this prevents the bacterial cell from growing. However, the mutation is occasionally advantageous and the cell is, for example, able to grow faster or survive longer than the other normal cells.

Gene Transfer Bacteria can transfer genetic material to other bacteria in three ways: conjugation, transduction, and transformation. Conjugation occurs during cell-to-cell contact and is somewhat similar to sexual recombination seen in other organisms. The transfered DNA can be either chromosomal or a small, circular piece of DNA called a plasmid.

Transduction is the transfer of genes from one bacterial cell to another by a bacterial virus. These viruses, called bacteriophages or phages, package a bacterial gene along with the viral genes and transfer it to a new cell.

A third method of transferring genes is transformation, which is also called DNA-mediated transformation. (The word *transformation* is sometimes used to define the change of normal animal cells to malignant cells—a completely different system.) In bacterial transformation, isolated DNA is mixed with viable cells. It then enters the cells, which are able to express these new genes. Although it would seem to be impossible for a large molecule such as DNA to enter through the cell wall and membrane of a living cell, this is indeed what happens.

Gene Regulation Another aspect of genetics is the expression of genes. A cell must be economical with its energy and material, and must not make enzymes or other products when they are not needed. On the other hand, a cell must be able to "turn on" genes when they are required in a particular environment. Gene regulation is examined in exercise 18.

Genetic study of bacteria and viruses is closely interwoven with the most general problems of their biology.

Joshua Lederberg

14, 15, 17, 18

<u>S</u> <u>E</u> <u>P</u> <u>U</u> <u>L</u> <u>C</u> <u>H</u> <u>U</u> <u>R</u> <u>A</u>
1 2 3 4 5 6 7 8 9 10

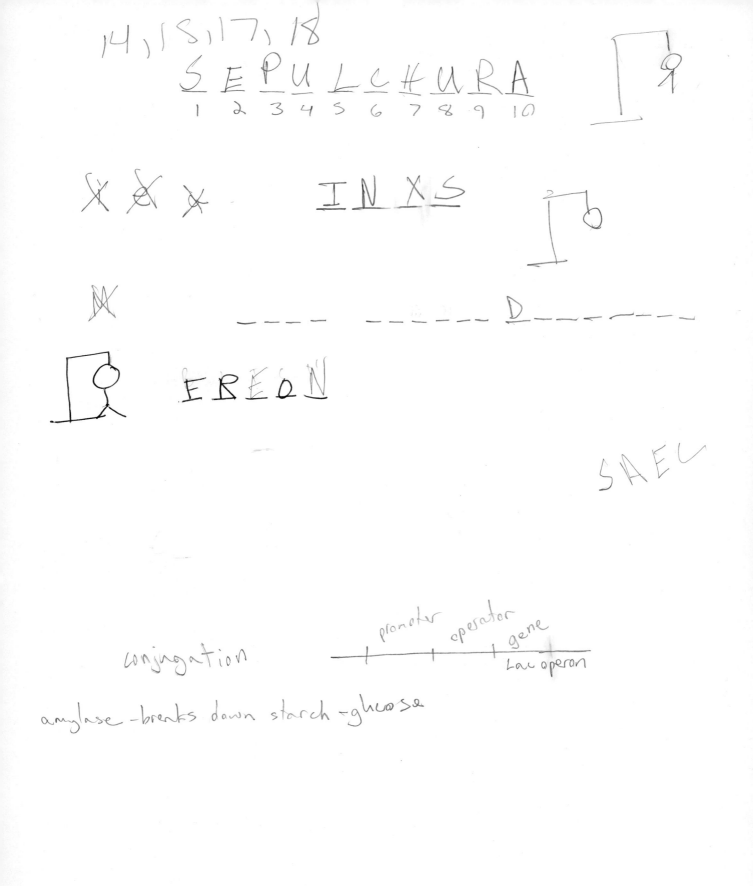

X X X   <u>I N X S</u>

M

____ ____ _____ _D_____ ____

EREON

SAEL

conjugation

promoter operator gene
|——————|————————|—————
Lac operon

amylase – breaks down starch – glucose

EXERCISE 15

Selection of Bacterial Mutants Resistant to Antibiotics

Getting Started

All the bacterial cells in a pure culture are derived from a single cell. These cells, however, are not identical because all genes tend to mutate and form mutant organisms. The spontaneous mutation rate of genes varies between 1 in 10^4 to 1 in 10^{12} divisions, and even though that is quite a rare event, significant mutations are observed because bacterial populations are very large. In a bacterial suspension of 10^9 cells/ml, one could expect 10 mutations of a gene that mutated 1 in every 10^8 divisions.

Mutant bacteria usually do not grow as well as the wild-type normal cell because most changes are harmful, or at least not helpful. If, however, conditions change in the environment and favor a mutant cell, it will be able to outcompete and outgrow the cells that do not have the advantageous mutation. It is important to understand that the mutation is a random event that the cell cannot direct. No matter how useful a mutation might be in a certain situation, it just happens to the cell, randomly conferring an advantage or disadvantage to it.

In this exercise, you select bacteria resistant to streptomycin. Streptomycin is an antibiotic that kills bacteria by acting on their ribosomes to prevent protein synthesis. (However, it does not stop protein synthesis in animals because eukaryotic ribosomes are larger than those of bacteria and therefore different.) Sensitive *E. coli* cells can become resistant to streptomycin with just one mutation.

In this exercise, you select organisms resistant to streptomycin by adding a large population of sensitive bacteria to a bottle of TS broth containing streptomycin. Theoretically, only organisms that already had a random mutation for streptomycin resistance will be able to survive and multiply (figure 15.1).

It is not possible in this exercise to test whether the streptomycin caused the mutations to streptomycin resistance or if the streptomycin selected

Figure 15.1 Selection of streptomycin-resistant *E. coli* cells.

A population of *E. coli* cells. Very few cells have mutated to streptomycin resistance.

—Sensitive cell
— Resistant cell

1 ml *E. coli*   1 ml *E. coli*

After incubation— sensitive cells

After incubation— resistant cells

TS broth   TS broth + streptomycin

mutants already existing in the population. Devising such an experiment was a very difficult problem for microbiologists in the 1940s. Can you propose one?

Definitions

Antibiotic. A substance produced by one organism, usually a microorganism, which kills or inhibits another organism.

Mutation. An inheritable change in the base sequences of DNA.

Mutation rate. The number of mutations per cell division.

Sensitive. An organism killed or inhibited by a particular antibiotic.

Wild type. A strain of bacteria as it is isolated from nature.

Objectives

1. To understand the concept of selection and its relationship to mutation.
2. To understand that mutations are random events, and that the cell cannot cause specific mutations to occur, no matter how advantageous they might be.
3. To count the number of streptomycin-resistant mutant bacteria that occur in an overnight culture of a sensitive strain.

References

Nester et al. *Microbiology: A human perspective*. Chapter 7, figure 7.12, 161–162; chapter 29, 21–28, 651.

Materials

Per team
First Session

Flasks (or bottles) containing 50 ml TS broth, 2

TS agar deeps, 2

Sterile Petri dishes, 2

1-ml pipets, 2

Overnight broth culture (~ 18 hours) of *Escherichia coli* (about 10^9 cells/ml)

Streptomycin solution at 30 mg/ml (about 1.5 ml/team)

Second Session

TS agar deeps, 2

Sterile Petri dishes, 2

1-ml pipets, 2

Tubes of 0.5 ml sterile water, 2

0.1-ml streptomycin

Procedure

First Session

1. Melt and place 2 TS agar deeps in a 50° C water bath.
2. Label one Petri plate and one flask "with streptomycin." Label the other flask and plate "without streptomycin control" (figure 15.2).
3. Add 0.3 ml streptomycin to the flask labeled streptomycin and 0.1 ml to one of the melted cooled agar deeps. Discard the pipet.
4. Immediately inoculate the agar deep with 1 ml of the bacterial culture, mix, and pour in the plate labeled "with streptomycin."
5. Add 1 ml of the bacteria to the tube of melted, cooled agar without streptomycin and pour into plate labeled "without streptomycin."
6. Add 1 ml of bacteria to each of the flasks.
7. Incubate the plates and flasks at 37° C. If using bottles lay them on their side to increase aeration.

Second Session

1. Melt and cool two tubes of TS agar in 50° C water bath.
2. Pour one tube of melted agar into a Petri dish labeled "without streptomycin" and let harden.
3. Add 0.1 ml streptomycin to the other tube of melted agar, pour into a Petri dish labeled "with streptomycin" and let harden.
4. Examine the bottles and plates inoculated last period. Note whether there is growth (turbidity) or not in both of the bottles. Count the number of colonies growing in the pour plates. How many streptomycin-resistant mutants/ml were present in the original inoculum? Compare it to the growth of organisms in the control plate without streptomycin. If there are more than 300 colonies or the plate is covered by confluent growth, record as TNTC or Too Numerous To Count. Record results.

Figure 15.2 Inoculating media with and without streptomycin with a culture of *E. coli* (Session 1).

1 ml *E. coli*   1 ml *E. coli*   1 ml *E. coli*   1 ml *E. coli*

TS broth
control

TS broth +
streptomycin

TS agar control   TS agar + streptomycin

Figure 15.3 Testing above incubated cultures for streptomycin sensitivity (Session 2).

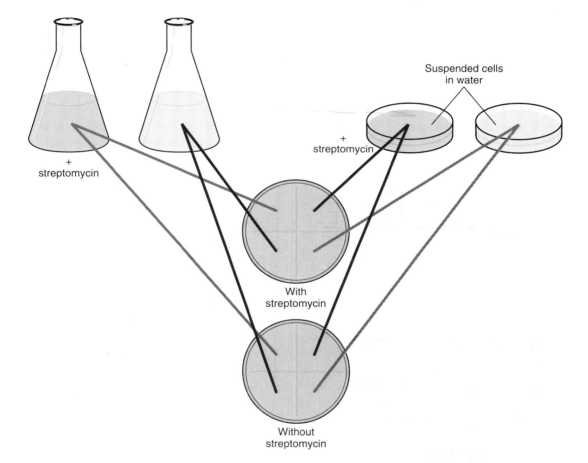

Suspended cells
in water

+
streptomycin

+
streptomycin

With
streptomycin

Without
streptomycin

5. Test the bacteria growing in the bottles and on the plates for sensitivity or resistance to streptomycin in the following way. Divide both agar plates in four sections as diagrammed in figure 15.3. Take a loopful of broth from the bottle without streptomycin and inoculate a sector of each agar plate. Do the same with the broth culture containing streptomycin.

6. Dig an isolated colony out of the agar plate containing the streptomycin and suspend it in a tube of sterile water. Use a loopful to inoculate the third sector of each plate. Also

STE

suspend some organisms from the control plate in saline (there will not be any isolated colonies) and inoculate the fourth sector. Incubate the plates at 37° C.

7. Predict which bacteria will be sensitive to streptomycin and which will be resistant.

Third Session

1. Observe growth on each sector of the plates and record results. Were they as you predicted?

2. Occasionally, mutants will not only be resistant to streptomycin, but also will require it. If you have one of these unusual mutants, be sure to show it to the instructor.

NIGHT RANGER

BJORK
1 2 3 4 5

CANDLEBOX

Inducible - lactose - only turned on when needed
B - galactosidase
glucose - galactose
ONPG - marker - turns yellow

catabolite repression
amylase - used to break down starch

EXERCISE

15

Laboratory Report: Selection of Bacterial Mutants
Resistant to Antibiotics

Results (after Third Session)

Source	Growth on TS Agar Plate	Growth on TS Agar Plate + Strp
TS broth (control)		
TS broth plus streptomycin		
TS agar (control)		
TS agar plus streptomycin		

How many organisms/ml were streptomycin resistant in the original overnight culture of sensitive *E. coli?*

Questions

1. Two bottles of TS broth (with and without streptomycin) were inoculated in Session 1 with 1 ml of an overnight culture of *E. coli.* After incubation, why was one population streptomycin sensitive and the other streptomycin resistant?

2. How were you able to estimate the number of streptomycin-resistant organisms already present in the overnight culture of *E. coli* growing in the TS broth?

3. Why should antibiotics not be used unless they are necessary?

EXERCISE

16

Transformation: A Form of Genetic Recombination

Getting Started

In this exercise, transformation is used to transfer the genes of one bacterium to another. It gives you a chance to see the results of what seems to be an impossible process—a huge DNA molecule entering an intact cell and permanently changing its genetic makeup.

Basically, the process involves mixing DNA from one strain of lysed (disrupted) cells with another strain of living cells. The DNA then enters the viable cells and is incorporated into the bacterial chromosome. The new DNA is expressed and the genetic capability of the cell may be changed.

In order to determine whether the bacteria are indeed taking up additional DNA, the two sets of organisms (DNA donors and DNA recipients) must differ in some way. One strain usually has a "marker" such as resistance to an antibiotic, or the inability to synthesize an amino acid or vitamin. In this exercise, a gene responsible for conferring resistance to the antibiotic streptomycin is transferred to cells that are sensitive to it (figure 16.1).

The organism used in this exercise is *Acinetobacter* (a sin NEET o bacter), a short, gram-negative rod found in soil and water. The prefix "a" means without, and "cine" means movement, as in cinema; thus *Acinetobacter* is nonmotile.

Definitions

Competent. Cells that are able to take up naked DNA. Some bacteria have to be in a certain physiological state or treated in special ways before they are competent, but the *Acinetobacter* strain used here is permanently competent.

DNase. An enzyme that cuts DNA, making it useless for transformation.

Figure 16.1 Transformation of cells with a gene conferring streptomycin resistance.

Cells resistant to streptomycin   Cells lysed, releasing DNA

DNA mixed with sensitive cells

Previously sensitive cells are now streptomycin resistant. They have been transformed.

Naked DNA. DNA released from lysed or disrupted cells and no longer protected by an intact cell.

Objectives

1. To understand the process of transformation and observe it in the laboratory.
2. To understand the use of genetic markers.
3. To understand the importance of controls in an experiment.

References

Nester et al. *Microbiology: A human perspective.* Chapter 7, 165–168.

Safety Precaution: *Acinetobacter* can cause pneumonia in immunologically compromised individuals.

Materials

Per team

TSY agar plate, 1

TSY agar plate with streptomycin (second session), 1

Broth culture of *Acinetobacter* Str^R (resistant to streptomycin), 1

Broth culture of *Acinetobacter* Str^S (sensitive to streptomycin), 1

Tube with 0.1 ml detergent SDS (sodium dodecyl sulfate) in 10× saline citrate, 1

Solution of DNase

1-ml pipet, 1

Class equipment
60° C water bath with test-tube rack

Procedure

First Session

1. Transfer 1.0 ml of Str^R *Acinetobacter* broth culture into the tube of SDS. Label the tube and incubate it in the 60° C water bath for thirty minutes. The detergent (SDS) will lyse the cells, releasing DNA and other cell contents. Any cells that are not lysed will be killed by the thirty-minute exposure to 60° C water. Label the tube DNA.

2. Divide the bottom of the TSY agar plate into 5 sectors using a marking pen. Label the sections DNA, Str^S (streptomycin sensitive), Str^R (streptomycin resistant), Str^S + DNA, and Str^S + DNase + DNA (figure 16.2).

3. Inoculate the plate as indicated by adding a loop of the broth culture, DNA, or DNase in an area about the size of a dime to each sector.
   a. DNA. The lysed mixture of Str^R cells is the source of DNA. It also contains RNA, proteins, and all the other cell components of the lysed cells, which do not interfere with the transformation. This mixture is inoculated onto the TSY agar plate to show that it contains no viable organisms. It is a control.

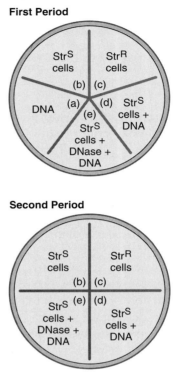

Figure 16.2 (*a–e*) Five labeled sectors of a TSY agar plate.

First Period

Second Period

TS agar + streptomycin

b. Str^S cells. Inoculate a loopful of the StrS culture. This step should demonstrate that the Str^S cells can grow on TSY agar (which is a control).

c. Str^R. Inoculate a loopful of the Str^R culture. This step should demonstrate that the Str^R culture is viable (it is a control).

d. Str^S cells + DNA. Inoculate a loopful of Str^S cells and add a loopful of the DNA (lysed Str^R cells) in the same area. THIS IS THE ACTUAL TRANSFORMATION. Str^R cells will grow here if transformed by the DNA.

e. Inoculate a loopful of Str^S cells as above, and in the same area add a loopful of DNase, then add a loopful of DNA. It is important to add these in the correct order (otherwise transformation will occur before the DNase can be added). This is a control to demonstrate that DNA is indeed the component of the lysed cells that is responsible for the transformation. The DNase should inactivate the DNA, preventing transformation.

4. Incubate the plates at room temperature for several days or at 37° C for 48 hours.

Second Session

1. Observe the plate you prepared in the first session. There should be growth in all sectors of the plate except the DNA sector (a). If the DNA control sector shows growth, it indicates that your crude DNA preparation was not sterile. If this has happened, discard your plates and borrow another student's plate after he or she is finished with it; there should be sufficient material for more than one team. Why is it so important that the DNA preparation is sterile?

2. Divide the bottom of a TSY + streptomycin plate into four sectors and label them StrS, Str^R, Str^S + DNA, and Str^S + DNase + DNA.

3. Streak a loopful of cells from the first plate to the corresponding sectors on the TSY + streptomycin plate. Lightly spread them in an area about the size of a dime. Cells growing on this plate must be streptomycin resistant.

4. Incubate at room temperature for several days or at 37° C for 48 hours or until cells have grown.

Third Session

Observe the TSY + streptomycin agar plate inoculated last period and record results. Did you transform the cells sensitive to streptomycin to cells that were resistant and could now grow on streptomycin?

NOTES:

EXERCISE

16

Laboratory Report: Transformation: A Form of Genetic Recombination

Results

Were *Acinetobacter* Str^S cells sensitive to streptomycin?

Were *Acinetobacter* Str^R cells resistant to streptomycin?

Was the DNA (cell lysate) free of viable cells?

Did transformation take place?

Did the DNase prevent transformation?

Yes/No	Sector that Demonstrates Observation

Questions

1. What two components were mixed together to show transformation?

2. What control showed that transformation and not conjugation or transduction was responsible for the results?

3. If the Str^S cells had grown on the TSY + streptomycin agar, would you have been able to determine if transformation had taken place? Explain.

4. If you had used a DNA lysate containing viable cells, would it have been possible to determine whether transformation had taken place? Explain.

5. How does transformation differ from conjugation and transduction?

EXERCISE

17

Bacterial Conjugation

Getting Started

Conjugation is one of the three mechanisms responsible for genetic transfer in bacteria (see Introduction). In conjugation, a cell of mating type F$^+$ attaches by a pilus to a cell of mating type F$^-$ and DNA is transferred from the F$^+$ cell to the F$^-$.

The donor cell is designated F$^+$ because it contains a fertility factor known as an F factor and the recipient is F$^-$ because it lacks the F factor. Some of the genes that make up the F factor code for the sex pilus, an appendage on the surface of the cell.

The F factor is usually located on a plasmid, which is a small, circular piece of DNA found in the cytoplasm of the cell. The genes found on plasmids are usually helpful to the cell in certain situations, but are not essential for the cell's normal metabolism. For example, the genes conferring a certain type of resistance to antibiotics are frequently carried on plasmids.

During conjugation, the pilus of the F$^+$ cell attaches to the F$^-$ cell and seems to be involved in bringing the cells into close contact. One strand of the plasmid enters the F$^-$ cell, the plasmid DNA replicates, and the F$^-$ cell becomes an F$^+$ cell. The cell can then express all the other genes (such as antibiotic resistance) contained on the plasmid. Conjugation is therefore a very important mechanism in the spread of antibiotic-resistant genes (figure 17.1a).

F factors can be responsible for the transfer of more than just the genes on a plasmid. The F factor can, on rare occasions, integrate into the bacterial chromosome, which can then be transferred during conjugation (see figure 17.1b). These strains are called Hfr strains for *high frequency of recombination.*

In this exercise, a donor Hfr strain containing the F factor in the chromosome is used to transfer chromosomal genes to an F$^-$ strain of *E. coli.* Each of the strains must have a genetic "marker" such as antibiotic resistance or inability to synthesize an essential cellular metabolite such as an amino acid or vitamin. In this case, the donor strain is unable to synthesize the amino acid methionine and the recipient cannot synthesize the amino acid threonine. Neither strain can grow on a mineral salts medium because it does not contain the amino acids that the strains require. (Mutants requiring a growth factor are termed auxotrophs.) If, however, the normal genes are transferred to the mutant recipient during conjugation, these cells then will be able to synthesize all the required amino acids. Recombinant bacteria will be able to grow on the mineral salts medium but the auxotrophic parent strains will not.

During conjugation, the chromosome is transferred in a linear manner with the F factor at the end—similar to a caboose. The longer the cells stay in contact, the more chromosome is transferred, but the cells usually break apart before the whole chromosome is transferred. It is important not to shake the culture during conjugation so that cells stay in contact as long as possible.

Definitions

Auxotroph.   An organism that cannot synthesize all its needed growth factors. The strain labeled meth- in the conjugation exercise is an auxotroph requiring methionine. It cannot grow on a mineral medium, but only on TS agar or another medium that contains methionine.

Conjugation.   A method of transferring DNA between bacteria requiring cell-to-cell contact.

F factor.   (F plasmid) Genes giving the cell the ability to transfer DNA via conjugation.

F$^-$.   Cells lacking the F factor, and are called the recipient bacteria.

F$^+$.   Cells containing the F factor, and are called donor bacteria.

Figure 17.1 (*a*) Conjugation: transfer of the F⁺ plasmid. (*b*) Conjugation: transfer of chromosomal DNA. From Eugene W. Nester, C. Evans Roberts, and Martha T. Nester, *Microbiology: A Human Perspective.* Copyright © 1995 Wm. C. Brown Communications, Inc., Dubuque, Iowa. All Rights Reserved. Reprinted by permission.

(a)

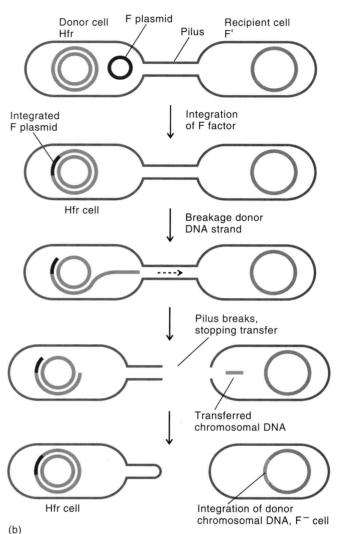

(b)

Hfr. (High frequency of recombination) F factor is incorporated in the chromosome of the bacteria, and consequently portions of the chromosome can be transferred to F⁻ bacteria.

Plasmid. A small circle of DNA found in some cells in addition to the chromosomal DNA.

Objectives

1. To increase knowledge of concepts and techniques used in the study of genetics.
2. To acquaint you with a laboratory method for demonstrating conjugation using auxotrophic organisms.
3. To demonstrate the importance of conjugation in transferring genes among bacteria.

References

Nester et al. *Microbiology: A human perspective.* Chapter 7, 168–171; figures 7.22, 7.24, and 7.26.

Materials

Per team

TS broth culture in log phase of *E. coli* A F⁺ methionine-(Hfr donor)

TS broth culture in log phase of *E. coli* B F⁻ threonine-(recipient)

TS agar plates, 1

Sterile test tube, 1

Sterile 9-ml water blanks, 5

Agar plates of mineral salts + 0.5% glucose (mineral medium or MM), 4

Sterile 1.0-ml pipets, 6

Sterile bent glass rod (sometimes called dally rods or hockey sticks), 1

Procedure

Steps are outlined in figure 17.2.

First Session

1. Label all tubes and plates as indicated in figure 17.2.
2. Transfer 1 ml of culture A into the water blank labeled A 1/10. With the same pipet, transfer a second 1 ml of culture A into the tube labeled A + B.
3. Transfer 1 ml of culture B into the water blank labeled B 1/10. With the same pipet, transfer a second 1 ml of culture B into the tube labeled A + B.
4. Gently mix the A + B tube and incubate it without shaking for 30 minutes at room temperature. This is the actual mating.
5. While the A + B mixture is incubating, inoculate half of the MM plate (labeled A) and half of the TS agar plate (labeled A) with loopfuls of culture A. Repeat with culture B. These are controls to verify that neither culture A nor B can grow on MM agar, but can grow on TS agar. These cultures were diluted 1/10 so that you would not carry over any of the rich TS broth medium when testing the organisms on the minimal medium.
6. After 30 minutes incubation, pipette 1 ml of the A + B mixture into the water blank labeled A + B 10^{-1}. Mix thoroughly and transfer 1 ml with a second sterile pipet to the water blank labeled A + B 10^{-2}. Again, mix thoroughly and transfer 1 ml with a third sterile pipet to the water blank labeled A + B 10^{-3}. Mix thoroughly.
7. With a sterile 1.0-ml pipet remove 0.1 ml from the A + B 10^{-3} dilution and add it to the surface of the MM plate labeled A + B 10^{-3}. Immediately spread the drop completely over the surface of the plate with the sterile bent glass rod.
8. Using the same pipet and bent glass rod, repeat the procedure to inoculate the MM agar plates labeled A + B 10^{-2} and A + B 10^{-1} from their respective dilutions. Remember, you can use the same pipet when going from low concentrations to higher concentrations.
9. Invert the agar plates and incubate at 37° C for 2–3 days.

Second Session

1. Examine the control MM and TS agar plates and record results. Are both cultures A and B viable auxotrophs?
2. Count the number of colonies on the plates that have countable numbers. Record the results.

Figure 17.2 Procedure for conjugation.

Conjugation

Controls

EXERCISE 17

Laboratory Report: Bacterial Conjugation

Results

Indicate where growth occurred.

	Culture A	Culture B
TS agar		
MM agar		
Culture an auxotroph?		

Plate Counts	10^{-1}	10^{-2}	10^{-3}
Plate labeled:			
Number of colonies			

1. How many recombinant organisms resulted from the mating? Show calculations.

2. If there were roughly 2×10^8 organisms in the A + B mixture, what percent of the original mixture resulted in recombinants?
Note: Since only the recipient cells can receive DNA and change genotype, the percent of recombinants should be based on the number of recipient cells, or 1×10^8 cells/ ml.

3. If the back mutation rate for threonine is about 1 in 10^8 cells, about how many mutants would you expect to be present per ml in the A + B mixture?

Questions

1. Why can the same pipet and bent glass rod be used to inoculate plates when starting with the most dilute mixture?

2. Why is an auxotrophic organism not able to grow on MM agar?

3. If either A or B could grow on the MM agar, how would that change the results?

4. Using the answers you calculated in number 2 in the Results section, how did the number of recombinants resulting from conjugation compare with the number you expect from back mutation?

5. Compare the effects of mutation and conjugation to produce organisms with new genetic capability.

NOTES:

EXERCISE

18

Gene Regulation: Induction and Catabolite Repression

Getting Started

A bacterial cell has all the genetic information to produce and operate a new cell. This includes the enzymes necessary for obtaining energy and synthesizing the necessary cellular components. A cell must work as efficiently as possible and carefully utilize the available nutrients without investing energy in enzymes not needed. Some enzymes are utilized in the basic energy pathways of the cell and are continually synthesized. These are called **constitutive enzymes.**

Others termed **inducible enzymes** are only needed when their specific substrate is available. For example, *E. coli* can break down the sugar lactose with an enzyme called *β*-galactosidase. If lactose were not present in the environment, it would be a waste of energy and of intermediate compounds to synthesize this enzyme. Therefore, these kinds of enzymes are called inducible enzymes because the presence of the substrate induces their synthesis.

How does the cell control inducible enzymes? In bacteria, this takes place on the level of transcription. Inducible enzymes can be found in an operon, which has a promoter and operator followed by the genes involved in the enzymatic activity (figure 18.1). The repressor binds to the operator gene blocking transcription if the substrate for the enzyme is not present.

When the substrate is present, it binds to the repressor and the RNA polymerase transcribes the genes.

Another method of control utilized by the cell is catabolite repression, which occurs when a cell has a choice of two sources of energy, one of which is more easily utilized than the other. An example is the presence of both glucose and starch. Glucose can immediately enter the glycolytic pathway, while starch must be first cleaved with amylase. Amylases and enzymes cost the cell energy and materials to produce, so therefore it is much more economical for the cell to utilize glucose if it is present. When glucose is present along with starch, the glucose represses the synthesis of amylase even though the enzyme would normally be induced in the presence of the starch.

The first exercise investigates induction of β-galactosidase, a particularly important enzyme system used in recombinant DNA techniques as a measure of gene expression.

The second exercise tests for the catabolite repression of amylase. Amylase is an exoenzyme that is excreted outside the cell because the large starch molecule may be too large to pass through the cell membranes.

Definitions

Operon.   A series of genes that is controlled by one operator (gene).

Substrate.   The molecule reacting with an enzyme.

Transcription.   Transfer of the genetic information from DNA to messenger RNA.

Objectives

1. To understand the concepts of induction and catabolite repression.
2. To understand the lac operon and the use of ONPG.
3. To understand the concept of catabolite repression and how it is tested.

References

Nester et al. *Microbiology: A human perspective.* Chapter 6, 147–149.

Figure 18.1 Inducible enzymes. (*a–b*) Regulation of enzyme synthesis in a degradative enzyme system. Note that the same genetic elements take part as in the regulation of a biosynthetic pathway. From Eugene W. Nester, C. Evans Roberts, and Martha T. Nester, *Microbiology: A Human Perspective.* Copyright © 1995 Wm. C. Brown Communications, Inc., Dubuque, Iowa. All Rights Reserved. Reprinted by permission.

Induction

Procedure

First Session

1. Inoculate each tube with a drop or loopful of
   E. coli.
2. Incubate at 37° C until the next laboratory
   period or at least 48 hours.

Second Session

1. Examine the tubes for growth—all the tubes
   should be turbid.
2. Add 1 ml of ONPG to each tube.
3. Incubate at room temperature for 30 minutes.
4. Examine the tubes to determine if the broth
   has turned yellow—an indication
   of the presence of the induced enzyme
   β-galactosidase.
5. Record the results.

Figure 18.2 Appearance of plates after flooding with
Gram's iodine.

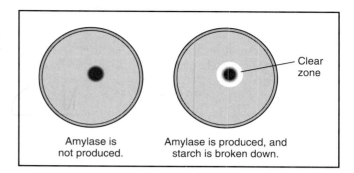

Amylase is
not produced.

Amylase is produced, and
starch is broken down.

Clear
zone

Catabolite Repression

Procedure

First Session

1. Label each plate.
2. Inoculate the middle of each plate with the
   Bacillus in an area about a few mm square.
3. Incubate at 30° C for a day or two. Try not to
   let the *Bacillus* grow over more than a third or
   half the plate. You may have to refrigerate the
   plates if the colony becomes too large.

Second Session

1. Flood the agar plate with Gram's iodine. The
   starch will turn purple. If the starch has been
   broken down with amylases, a clear zone will
   appear around the colony (figure 18.2).
2. Record the results.

NOTES:

Physical

enzymes – inactivated, denatured → coagulate

veg. cells 60°–70°C 10min

spores > 100°C   Fungus 50°–60°C 10min

                        Spores 70°–80°C 10min

Pasteurization

high short 71°C for 15sec.

low long 63°C for 30min

Tyndallization

3 times { 100°C – 20min

           24hrs – 37°C

dehydration – enzyme

2 hours – 165°C

Thermal Death Point (TDP)
Lowest Temp needed to
kill *all* organisms in 10min.

10

20

30

40

EXERCISE

Laboratory Report: Gene Regulation: Induction and Catabolite Repression

Results

	Glucose	Lactose	Glycerol
Color			
β-galactosidase present?			

Results of catabolite repression

	Starch	Starch + Glucose
Zone of clearing present		
Presence of amylase		

Questions

1. Which substrate induced β-galactosidase?

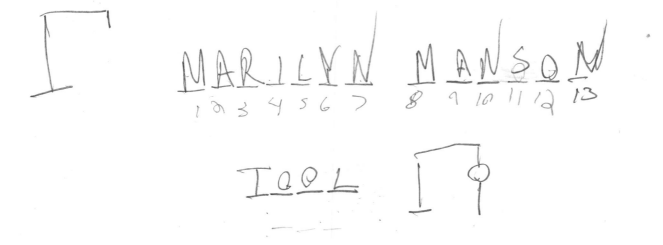

2. What results would you expect if β-galactosidase were a constitutive enzyme?

3. In the catabolite repression exercise, did the *Bacillus* have the capacity to synthesize amylase (amylase +)?

4. Did you observe catabolite repression when glucose was added to the starch?

5. What results would you expect if amylase were a constitutive enzyme?

PART TWO

The Other Microbial World

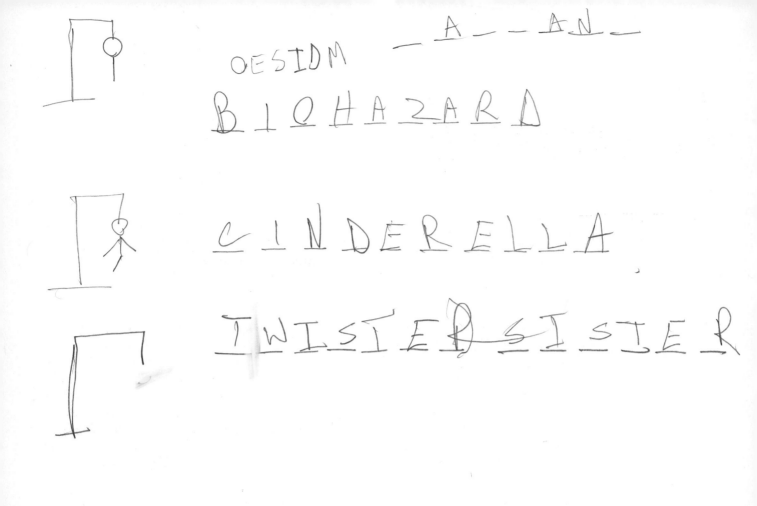

OESIDM _ _ A _ _ A N _

BIOHAZARD

CINDERELLA

TWISTED SISTER

*I*NTRODUCTION to the Other Microbial World

The phrase, the "other microbial world" refers to organisms other than bacteria, the major organisms of study in other parts of the manual. All of the organisms included here, with the exception of viruses, are eukaryotic organisms, many of which are of medical importance. Included are members of the non-filamentous fungi (primarily yeasts), the filamentous fungi (molds), and intestinal animal parasites of medical importance, the protozoa and helminths. Viruses that infect both prokaryotes (bacteria) and eukaryotes (animal and plant cells) are introduced; but for reasons to be explained later only bacterial and perhaps plant viruses will be made available for laboratory study.

Mycology, or the study of fungi, is the subject of exercises 19 and 20. The word *fungus* is probably derived from the Greek word "spongus" meaning a spongy or filamentous growth. Included in the fungi are a wide variety of forms, ranging from unicellular microscopic cells, such as yeast, to filamentous multicellular structures. Many of the latter forms such as mushrooms, puffballs, toadstools, and bracket fungi are visible with the naked eye.

Most of the yeasts and filamentous fungi found in nature are nonpathogenic. In fact, many are important contributors to our well-being; for example, the yeast *Saccharomyces* is important in the manufacture of bread, beer, and wine; and the filamentous fungus *Penicillium chrysogenum* produces the antibiotic penicillin. Some are also associated with various kinds of spoilage (moldy jam and bread, mildew of clothing, and etching of glass lenses in warm, humid environments) and, more recently, with the production of food toxins known as aflatoxins by certain species of the genus *Aspergillus*.

Fungi are eukaryotic organisms that differ from algae in that they lack chlorophyll. Fungi differ from bacteria in that the cells are much larger and contain membrane-bound organelles. In addition, bacteria are prokaryotes.

The latter difference, in turn, affects antibiotic therapy; for example, antibiotics effective against prokaryotes are often ineffective against eukaryotes. If an antibiotic is effective against a fungus it may also, depending on the mode of action of the antibiotic, damage the human host since humans are eukaryotic. Consequently, antibiotic control of fungal infections is usually more difficult than control of bacterial infections. Fortunately, many fungal infections are opportunistic* infections so that healthy individuals rarely acquire them other than, perhaps, cutaneous fungal infections such as athlete's foot. Nevertheless, there are a number of important mycotic diseases with which you should familiarize yourself.

Fungi can be cultivated in the laboratory in the same manner as bacteria. Physiologically, all fungi are heterotrophs (they require an organic source of carbon, such as glucose) and most are aerobic, although some are facultative, that is, able to carry out both aerobic and anaerobic metabolism. Most fungi grow best at temperatures of 20 to 30° C, although some grow well at temperatures as high as 45 to 50° C (such as *Aspergillus fumigatus*, an opportunistic filamentous fungus known to cause pulmonary aspergillosis).

Parasitic diseases constitute a major worldwide public health problem. In developed countries such as the United States, diagnosis becomes essential in patients with unexplained malnutrition, malabsorption, immune dysfunction, or gastrointestinal disturbances (Galland and Bueno, 1989). In developing countries, parasitic diseases are prominent due to poverty, malnutrition, lack of sanitation, and lack of education. A simple family survey for intestinal parasites in a small Central American village revealed that every family member harbored at least three types of parasites. Effects of human

*Opportunistic infections are associated with debilitating diseases (such as cancer) and use of cytotoxic drugs, broad-spectrum antibiotics, and radiation therapy, all of which can suppress the normal immune response.

Figure I.7.1 Tobacco mosaic virus.Electron micrograph (× approximately 70,000). Compare length and width with a rod-shaped bacterium.
© Omikron/Photo Researchers, Inc.

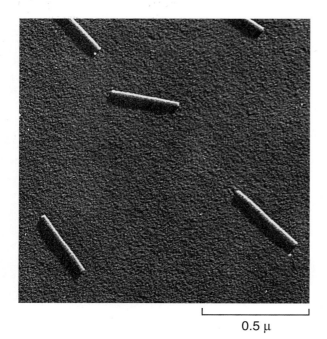

0.5 μ

parasitic disease range from minimal, with some nutritional loss but only minor discomfort (common in pinworm and *Ascaris* infections), to debilitating and life-threatening disorders such as malaria and schistosomiasis. Malaria has probably caused more suffering, impoverishment, and death to more people over thousands of years than any other infectious disease.

Fecal contamination of drinking water by wild animal carriers such as beavers has in recent years caused major outbreaks of **giardiasis** (an intestinal disease) in various parts of the United States. The causative agent, the protozoan *Giardia lamblia* (see figure 34.3), produces **cysts** that are quite small, thereby enabling passage sometimes through faulty water supply filters. The cysts often resist chlorination. The protozoan flagellate, *Trichomonas vaginalis*, is a common cause of vaginitis in women and is often sexually transmitted. Pinworm infestations are a problem among elementary school children. About 20% of infections in domestic animals are caused by protozoan and helminthic (worm) agents.

Because of their worldwide public health importance and their natural history differences from bacteria and fungi, we believe protozoa and helminths merit their own laboratory session (see exercise 21). It provides insights on the diseases they cause and the techniques used for their diagnosis and identification.

Virology, the study of viruses (the word for poison in Greek), also had early roots, although somewhat mysterious. Mysterious, in part, because the viruses could not be seen, even with a light microscope, and yet when sap from the leaf of an infected tobacco plant was passed through a filter that retained bacteria and fungi, the clear filtrate retained its infectious properties. Two early investigators, Iwanowski in 1892 and Beijerinck in 1898, described various properties of the fluid. Such properties included the infectivity of filtered sap, the need for actively growing plants to induce infection, sap stability to drying and cold temperatures, and sap inactivation by heating. Iwanowski concluded that the filtered sap contained a very small bacterium. Beijerinck concluded from his studies that the tobacco mosaic disease was not caused by microbes but by a living liquid virus, which he also described as a "contagium vivum fluidium."* It was not until the mid 1930s that viruses were first observed with the advent and aid of the electron microscope. In 1935, Wendell Stanley succeeded in crystallizing Tobacco Mosaic Virus (TMV), enabling him to observe that it was structurally different from living cells (figure I.7.1).

You may have an opportunity to repeat and extend some of Beijerinck's studies with TMV (see exercise 22, Notes to Instructor).

Bacterial viruses known as **bacteriophage** were first described by Twort (1915) and later by d'Herelle (1917). d'Herelle observed their filterable nature and their ability to form **plaques** on an agar plate seeded with a lawn of the host bacterium (see figure 22.2). Both Twort and d'Herelle worked with coliform bacteria isolated from the intestinal tract.

The discovery of plaque formation on an agar medium was a godsend for future virology research in that it provided a fast, easy way to recognize, identify, and quantify bacterial virus infections.

---

*A likely translation is "a contagious living fluid that is soluble."

I–2   Introduction to the Other Microbial World

Figure I.7.2 Mammalian virus plaques showing different cytopathic effects. Photographed 5 days after infection of the growth medium (a single layer of monkey kidney cells) with the various mammalian viruses. When in monolayers, the viruses are able to form plaques (a form of CPE), which can be detected macroscopically. From *Diagnostic Procedures for Viral, Rickettsial and Chlamydial Infections*, 5th Edition. Copyright © 1979 by the American Public Health Association. Reprinted with permission.

Viruses that attack mammalian cells also form structures analogous to plaques when cultivated on growth media able to support mammalian cell growth. Rather than plaques, the structures formed are described as **cytopathic effects (CPE)** (figure I.7.2). The CPE observed is dependent upon the nature of both the host cell and invading virus. The same can be said about the nature of bacterial plaque formation.

The advantage of using a bacteriophage to demonstrate virus growth is the ease of culturing the host cells. Mammalian cells require complex growth media including blood serum as well as prolonged incubation with the virus before being able to observe CPE. Therefore, we will work with an *E. coli* bacteriophage that can either be isolated and concentrated from sewage or obtained from a pure culture collection (see exercise 22).

Having made sundry efforts, from time to time, to discover, if 'twere possible, the cause of the hotness or power whereby pepper affects the tongue (more especially because we find that even though pepper hath lain a whole year in vinegar, it yet retaineth its pungency); I did now place anew about ⅓ ounce of whole pepper in water, and set it in my closet, with no other design than to soften the pepper, that I could the better study it. This pepper having lain about three weeks in the water, and on two several occasions snow-water having been added thereto, because the water had evaporated away; by chance observing this water on the 24th of April, 1676, I saw therein, with great wonder, incredibly many very little animalcules, of divers sorts; and among others, some that were 3 or 4 times as long as broad, though their whole thickness was not, in my judgment, much thicker than one of the hairs wherewith the body of a louse is beset. . . . The second sort of animalcules consisted of a perfect oval. They had no less nimble a motion than the animalcules first described, but they were in much greater numbers. And there was also a third sort, which exceeded both the former sorts in number. These were little animals with tails, like those that I've said were in rainwater.

The fourth sort of little animals, which drifted among the three sorts aforesaid, were incredibly small; nay, so small, in my sight, that I judged that even if 100 of these very wee animals lay stretched out one against another, they could not reach to the length of a grain of coarse sand; and if this be true, then ten thousand of these living creatures could scarce equal the bulk of a coarse sand-grain.

I discovered yet a fifth sort, which had about the thickness of the last-said animalcules, but which were near twice as long.

DOBELL, *Antony van Leeuwenhoek and his Little Animals*

conidia - naked ○ ⚇

arthrospores - ⬡⬡⬡⬡⬡ ⬡⬡

chlamydospores - ◉

blastospores - ∞

most bacteria reproduce asexually

Glucose, lactose, maltose

Lacto Phenol cotton blue

Candida albicans
Saccharomyces cerevisiae

Figure 19.1 Drawings of vegetative cell morphology of some representative yeasts. (*a*) *Saccharomyces cerevisiae;* (*b*) *Candida* sp.; (*c*) *Selenotila intestinalis;* (*d*) *Trigonopsis variablis.*

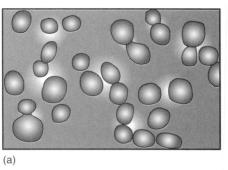

(a)

(b)

(c)

(d)

Getting Started

Yeasts, like molds, are fungi, but unlike molds they are usually unicellular rather than filamentous. Those that appear filamentous, such as *Candida* species, form a false filament (**pseudohyphae,** see figure 19.3), which often reverts to unicells with a change in growth environment (figure 19.1*b*). The latter phenomenon is an example of **yeast dimorphism.**

Yeast cell morphology varies. Examples of cell shapes include oval (*Saccharomyces cerevisiae*), lenticular (*Selenotila intestinalis*), triangular (*Trigonopsis* species), pear-shaped (*Pichia* species), and cigar-shaped (*Schizosaccharomyces* species) (see figure 19.1). Their mode of asexual reproduction is usually by means of **budding,** a process whereby an outgrowth develops from the **mother cell** and sepa-

rates from it upon reaching maturity, forming a new cell called the **daughter cell.** A few yeasts, such as *Schizosaccharomyces* species, reproduce asexually by **fission,** the same process as bacteria. A change in yeast cell morphology can also elicit a change in yeast colonial morphology.

Under certain environmental conditions many yeasts reproduce sexually, forming naked (not enclosed by a fruiting body) **asci** that usually contain one to four **ascospores** (figure 19.2). Due to their mode of sexual reproduction, yeasts are members of the fungal class Ascomycetes.

Classification or taxonomy of yeasts usually requires a study of both yeast physiology and morphology. However, they can sometimes be identified to the genus level solely by a study of their morphology. One of the most widely used physiology tests determines the ability of a particular yeast to

Figure 19.2 Asci and ascospores of *Saccharomyces cerevisiae* showing asci with four or fewer spores. Courtesy of University of Washington.

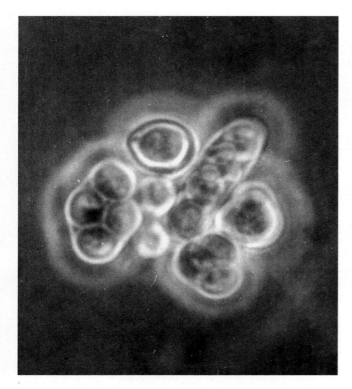

assimilate and/or **ferment** a series of growth media containing different sugars as the sole carbohydrate source. Assimilation studies are conducted by either the **auxanographic** method in which agar plates seeded with the test yeast are inoculated on the surface with paper discs containing the various test sugars, or by the tube method in which test tubes containing the various test sugars are inoculated with a drop of yeast suspension. The fermentation tests are conducted in tubes of broth containing an inverted tube (or Durham tube) to detect gas production. In the assimilation tests, the presence of yeast growth (a colony on a solid medium or sediment in a liquid broth medium) constitutes a positive assimilation test. In contrast, in the fermentation test, the presence of gas in the Durham tube constitutes a positive fermentation test. The presence of sediment or a change in color of the pH indicator are not indicative of fermentation.

The assimilation and fermentation tests are good examples of the oxidative and glycolytic metabolic pathways, respectively.

A description of the more important pathogenic yeasts is described in table 19.1. Note that two of these yeasts are opportunistic (*Cryptococcus* and *Candida*). *Candida albicans* is an inhabitant of the normal flora, and is found in the mouth, intestinal tract, and vagina. *Candida* infections often occur when the body is immunosuppressed by either cytotoxic drugs or radiation therapy. Several of the seven remaining *Candida* species are also pathogenic (Al-Doory, 1980).

Morphological characteristics used to differentiate *Candida albicans* from nonpathogenic species include:

1. Formation of a creamy colony on Sabouraud's dextrose agar.
2. Formation of pseudohyphae and budding yeast cells termed **blastospores** on Sabouraud's and cornmeal agars (figure 19.3).
3. Formation of thick-walled, resistant reproductive structures called **chlamydospores** on cornmeal agar (see figure 19.3).
4. Formation of **germ tubes** when suspended in either egg white or serum for a few hours (figure 19.4).

Definitions

Ascospore. Sexual spore characteristic of the fungus class Ascomycetes.

Ascus. Saclike structure containing ascospores formed by Ascomycetes during sexual reproduction.

Assimilation. Ability to utilize carbohydrates for growth in the presence of oxygen. (See oxidative pathway.)

Auxanographic. A method for determining the effect of a particular agent or condition on the growth of a microorganism.

Blastospores. Asexual spore formed by budding from a cell or from a hyphae.

Budding. Asexual process of reproduction in which a daughter cell (bud) evolves from either a larger cell (mother cell) or from a hyphae.

Table 19.1   Some Important Pathogenic Yeasts or Yeastlike (Dimorphic) Organisms

Organism	Morphology	Ecology & Epidemiology	Diseases	Treatment
Cryptococcus neoformans	Single budding cells, encapsulated	Found in soil and pigeons' nests. No transmission between humans and animals. May be opportunistic.	Meningitis, pneumonia, skin infections, visceral organs	Amphotericin B
Candida albicans	Budding cells, pseudomycelium formation chlamydospores	Normal inhabitants of mouth, intestinal tract. Opportunistic infections.	Thrush, vaginitis, nails, eyes, lungs, systemic infections	Alkaline mouth and douche washes; parahydroxy-benzoic acid esters; amphotericin B
Blastomyces dermatitidis	37° C: single large budding cells, 20° C: mold with conidia	Disease of North America and Africa Found occasionally in nature. No transmission between humans and animals.	Primarily lungs Also skin and bones	High-calorie, high-vitamin diet; bed rest; aromatic diamidines; amphotericin B
Paracoccidioides braziliensis (*Blastomyces braziliensis*)	37° C: single and multiple budding cells 20° C: mold with white aerial mycelium	Confined to South America. Workers in close association with farming.	Chronic granulomatous infection of mucous membranes of mouth, adjacent skin, lymph nodes, viscera	Sulfonamides; amphotericin B
Histoplasma capsulatum	37° C: single small budding cells 20° C: mold with tuberculate chlamydospores	Saprophyte in soil. No transmission between humans and animals. Epidemics from silos, chicken houses, caves, etc.	Primarily lungs, may spread to reticuloendothelial system	Amphotericin B
Coccidioides immitis	37° C: thick-walled endospore filled, spherical cells 20° C: mold with arthrospores	Disease primarily of arid regions, such as San Joaquin valley Dust-borne disease. No transmission between humans and animals.	Primarily lungs may disseminate particularly in African-Americans and is highly fatal	Bed rest; amphotericin B; surgery for lung lesions

Figure 19.3 Chlamydospores and smaller blastospores, (attached to pseudohyphae) of *Candida albicans* grown on cornmeal agar. Preparation stained with methylene blue.
Courtesy of the Upjohn Co.

Figure 19.4 Germ tubes formed by *Candida albicans* grown on egg albumin. Phase contrast magnification.
Courtesy of the Upjohn Co.

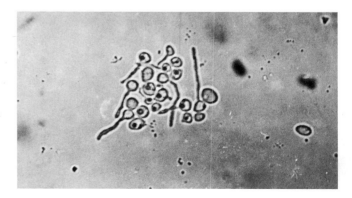

Chlamydospore. A thick-walled, resistant asexual spore formed either between cells of a hyphae or terminally.

Daughter cell. A new yeast cell. (See also budding.)

Fermentation. A growth process occurring in the absence of oxygen in which the final electron acceptor is an organic compound.

Fission. An asexual reproductive process in which one cell splits into two or more independent daughter cells.

Germ tube. A tubelike outgrowth from an asexual cell that develops into a hyphae.

Glycolytic pathway. The initial sequence by which carbohydrates are degraded, often called the Embden-Meyerhoff pathway. It involves a series of fermentation steps.

Hypha (hyphae). Threadlike filament(s) that form a mat described as mycelium.

Mother cell. The parent of the daughter cell(s).

Opportunist. An organism that causes disease only in hosts with impaired defense mechanisms.

Oxidative pathway. The oxidation of carbohydrates in the presence of oxygen in which the final electron acceptor is oxygen. Commonly known as the tricarboxylic acid, or Krebs, cycle.

Pseudohyphae. Structures resembling hyphae that are formed by yeastlike fungi.

Yeast dimorphism. Existing in two growth forms, such as the mold phase (formation of pseudohyphae) and yeast phase (single cells).

Objectives

1. To introduce information about some of the important characteristics of yeasts, and methods used for identification of both pathogenic and nonpathogenic yeasts.
2. To allow you to evaluate some of the methods used to identify both a nonpathogenic nonopportunist yeast (for example, *Saccharomyces cerevisiae*) and a pathogenic

opportunist yeast (for example, *Candida albicans*). Methods for study include colonial and vegetative cell morphology, sugar fermentations, sexual spore formation, chlamydospore and pseudohyphae formation, and germ tube production.

References

Al-Doory, Y. *Laboratory medical mycology.* Philadelphia: Lea and Febiger, 1980.

Barnett, J. A.; Payne, R. W.; and Yarrow, D. *Yeasts: Characteristics and identification,* 2d ed. New York: Cambridge University Press, 1991.

Nester et al. *Microbiology: A human perspective.* Chapter 5, pp. 111–112; chapter 12, pp. 274–280; and chapter 33, pp. 730–733.

Phaff, H. H.; Miller, M. W.; and Mrak, E. M. *The life of yeasts,* 2d ed. Cambridge, Mass: Harvard University Press, 1978. This book provides an ecological approach to the study of yeast, including where to find different yeasts in their natural habitats.

Materials

Per team of two students

Cultures

Sabouraud's dextrose broth cultures (48 hour, 25° to 30° C) of *Saccharomyces cerevisiae* and *Candida albicans*, 1 of each

Sabouraud's dextrose agar slant culture (48 hour, 25° to 30° C) of *C. albicans*

Sabouraud's dextrose agar, 1 plate

Glucose-acetate agar, 1 plate

Cornmeal agar, 1 plate

Glucose, maltose, and lactose broth containing Durham tubes, 2 tubes of each broth

12 × 75-mm tube containing either 0.5 ml of serum or raw, nonsterile egg white, 1

Sterile Pasteur pipets, 4

Ruler divided in mm

Dropping bottle containing methylene blue

Procedure

First Session

Suspend the broth cultures of *Saccharomyces cerevisiae* and *Candida albicans*.

1. Fermentation Study. Inoculate each of the carbohydrate fermentation tubes (glucose, maltose, and lactose) with a *loopful* of the respective test cultures. Place the properly labeled tubes in a container and incubate at 25° to 30° C for 48 hours.

2. Colonial and Vegetative Cell Morphology Study. Divide the bottom surface of the Sabouraud's dextrose agar plate in half with a marking pencil. Label one half *S.c.* for *Saccharomyces cerevisiae* and the other half *C.a.* for *Candida albicans*. With a Pasteur pipet, inoculate the agar surface of each sector with a small drop of the respective test yeast. Allow the inoculum to soak into the agar before incubating right side up in the 25° to 30° C incubator for 48 hours.

3. Sexual Sporulation Study. Inoculate the center of the glucose-acetate agar plate with a drop of the *S. cerevisiae* broth culture inoculum. Allow the inoculum to soak into the agar before incubating right side up in the 25° to 30° C incubator for 48 hours.

4. Chlamydospore Formation Study. Inoculate the center of the cornmeal agar plate with a drop of the *C. albicans* broth culture inoculum. With a sterile inoculating *needle*, streak the drop across the length of the plate two times with just enough pressure to mark but not cut the agar. Next, streak back and forth across the marked area. Grasp a coverslip with a pair of sterile tweezers (sterilize by dipping in 95% alcohol and passing through the Bunsen burner flame) and then place the coverslip over a portion of the streaks. Incubate the plate right side up in the 25° to 30° C incubator for 48 hours.

5. Germ Tube Formation Study.
   a. Preparation. Remove a loopful of the *Candida albicans* culture from the surface of the Sabouraud's dextrose agar slant culture. Transfer and emulsify the inoculum in the tube of serum or raw egg white. Incubate the tube for 2½ to 3 hours in the 37° C incubator.
   b. Observation. Mix the serum or raw egg white with a Pasteur pipet; using a single drop of the mixed suspension, prepare a wet mount slide preparation. Examine first with the low power objective and next with the high power objective. Check for short germ tubes that give the cells a drumstick appearance (see figure 19.4). Prepare drawings of your findings in the Laboratory Report.

Note: If insufficient time remains for observation, the tubes can be held for observation until the second session by storing them in a covered container in the refrigerator.

Second Session

1. Fermentation Study. Examine the fermentation tubes for the following and record your results in table 19.2 of the Laboratory Report:
   a. Presence or absence (+ or −) of turbidity (growth).
   b. Presence or absence (+ or −) of gas in the inverted Durham tube.
   c. Change in color of the pH indicator dye. A change to a yellow color is indicative of acid production.

Note: Gas production is indicative of fermentation (glycolysis). To detect false negative results caused by supersaturation of the broth, all tubes giving an acid reaction should be shaken lightly and the cap vented. This operation is frequently followed by a rapid release of gas. All positive fermentation reactions with a carbohydrate are accompanied by positive assimilation of that carbohydrate; however, sugars may be assimilated without being fermented.

2. Colonial and Vegetative Cell Morphology Study.
   a. Colony characteristics. If possible, observe the Sabouraud's agar plate over a 5- to 7-day incubation period. Make note of the following in table 19.3 of the Laboratory Report: Colony color; consistency (soft, firm), probe the colony

with a needle for this determination; colony diameter (mm); appearance of colony surface (rough or smooth, flat or raised); and appearance of colony edge (circular or indented).

b. Vegetative cell morphology. Remove a loopful of surface growth from each colony and prepare wet mounts. Observe with the high dry objective, noting the shape and size of the cells and the presence or absence of pseudohyphae (see figure 19.3). Prepare drawings of your findings in the Laboratory Report.

3. Sexual Sporulation Study. With a sterile loop, touch the *S. cerevisiae* colony on the glucose-acetate agar plate and prepare a wet mount. Observe with the high dry objective and look for the presence of asci containing 1 to 4 ascospores (see figure 19.2). Prepare drawings of your findings in the Laboratory Report.

Note: In the event that you do not find asci, reincubate the plate for up to one week, and re-examine periodically. Some yeast strains take longer than others to produce sexual spores.

4. Chlamydospore Formation Study. Remove the cover of the cornmeal agar plate and place the whole plate on the microscope stage; with the 10× objective focus on the edge of the coverslip and search for chlamydospores that are quite large (7 to 17 μm). When present, they will usually be found underneath the coverslip near the edges. Prepare a drawing of your findings in the Laboratory Report (see figure 19.3). You can also remove some of the agar with a loop from the marked area, place it on a slide, and prepare a wet mount with a drop of methylene blue, a dye capable of staining the chlamydospores. Observe with low and high dry objectives.

Aspergillus

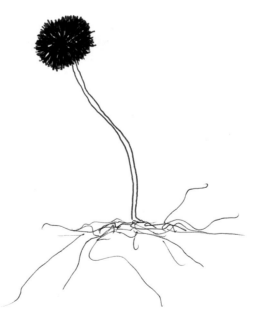

Plate 1 Appearance of bacterial colonies growing on agar. © Larry Jensen/Visuals Unlimited

Plate 2 Appearance of mold (*Penicillium*) growing on an agar plate. © Raymond B. Otero/Visuals Unlimited

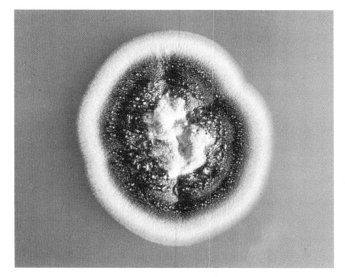

Plate 3 Bacterial shape. *Staphylococcus aureus* cocci as seen with the bright-field light microscope (×1,000). © LeBeau/Biological Photo Service

Plate 4 Bacterial shape. *Bacillus megaterium* rods or bacilli as seen with the bright-field light microscope (×600). © George Wilder/Visuals Unlimited

Plate 5   Bacterial shape. *Rhodospirillum rubrum* as seen with the bright-field light microscope (×500). © Thomas Tottelben/Tottelben Scientific Co.

Plate 6   Budding yeast photographed with bright-field (left) and dark-field (right) microscopy. © Dr. Edward Bottone, The Mount Sinai Hospital, New York, New York

Plate 7   Light micrograph (×900) of a Gram-stained mixture of gram-positive *Staphylococcus aureus* (purple cocci) and gram-negative *Escherichia coli* (pink rods). Courtesy of John Harley

Plate 8   Macroscopic fungi. (*a*) *Calvatia gigantea,* one of the largest visible fungi, which is estimated to produce as many as 26 million reproductive spores; (*b*) *Polyporus arcularius,* a large visible bracket fungus growing on a decaying tree trunk. (a) © Glenn M. Oliver/Visuals Unlimited (b) ©Dick Poe/Visuals Unlimited

(a)

(b)

Plate 9   Green mold (probably a *Penicillium* sp.) prevalent on stored fruits. The white areas consist largely of fungus mycelium, which removes nutrients from the strawberry, and eventually forms asexual reproductive structures containing spores (conidia), which when released by air currents further spread the infection (green area on right bottom of strawberry). © Matt Meadows/Peter Arnold, Inc.

Plate 10   Ringworm lesions on the scalp. Note hair loss and scaling of scalp. © Everett S. Beneke/Visuals Unlimited

Plate 11   Fixed cutaneous sporotrichosis. Identification is based on the type of lesion formed and culture characteristics (see also color plates 12, 13, and 14). © Everett S. Beneke/Visuals Unlimited

Plates 12, 13, and 14 *Sporotrichum schenkii,* a dimorphic fungus that forms a moldlike colony with hyphae containing terminal clusters of pyriform conidia and mycelia when incubated at 20° C (plates 12 and 13) and fusiform to round yeast cells when incubated at 37° C (plate 14). Courtesy of the Upjohn Co.

Plate 13

Plate 14

Plates 15, 16, and 17 *Coccidioides immitis,* a dimorphic fungus that forms a white, moldlike colony containing septate hyphae and chains of thick-walled arthrospores when grown on an agar medium at 20 to 37° C (plates 15 and 16), and round, thick-walled spherules (20–80 μm in diameter) containing many small endospores (2–5 μm in diameter) when isolated from sputum, pus, gastric contents, or spinal fluid (plate 17). Courtesy of the Upjohn Co.

Plate 16

Plate 17

Plate 18   *Paramecium* from pond water showing cilia and internal structures. Phase-contrast microscopy (×100).
© Mike Abbey/Visuals Unlimited

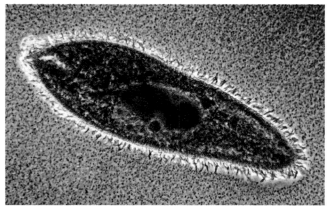

Plate 19   *Schistosoma miracidium,* or snail phase. Note the ciliated larvae.© Cabisco/Visuals Unlimited

Plate 20   Sexual mating of male and female *Schistosoma* worms occurs in the liver. The male worm holds the female in a groove located on his ventral surface.
© Science VU/Visuals Unlimited

Plate 21   Carbohydrate fermentation. © Raymond B. Otero/Visuals Unlimited

Plate 22   Rapid bacterial identification system, Enterotube II. Uninoculated tube (top) and inoculated tubes (bottom). Courtesy of the University of Washington

Plate 23   Appearance of normal human peripheral blood. Note the frequency of white blood cells as compared to red blood cells. In abnormal blood, the number of white blood cells would increase. Courtesy of the University of Washington Photo Library

Plate 24   Photograph of a line of identity on a *Candida* immunodiffusion agar plate.

Plate 25   Characteristics of *Staphylococcus aureus* colonies when cultivated on mannitol salts agar. Note yellow color because of acid production.

Plate 26   Colony characteristics of *E. coli* on an Endo MF medium. Note the pink to dark red colonies, some of which appear to have a golden green metallic sheen.

EXERCISE

Laboratory Report: Nonfilamentous Fungi: Yeasts and Yeastlike Fungi

Results

1. Fermentation Study. Examine tubes and record results (+ or −) in table 19.2.

Table 19.2 Fermentation Activity with Different Carbohydrates

Yeast Strain	GLUCOSE			MALTOSE			LACTOSE		
	Cloudy	Gas	Acid	Cloudy	Gas	Acid	Cloudy	Gas	Acid
C. albicans									
S. cerevisiae									

2. Colonial and Vegetative Cell Morphology Study
   a. Colony characteristics (see Procedure in Second Session step 2a) and enter results in table 19.3.

Table 19.3 *C. albicans* and *S. cerevisiae* Colonial Characteristics

Yeast Strain	COLONIAL MORPHOLOGY				
	Colony Color	Consistency	Diameter (mm)	Surface Appearance	Edge Appearance
Candida albicans					
Saccharomyces cerevisiae					

b. Vegetative cell morphology (see Procedure in Second Session step 2.b) and enter results below:

Candida albicans

Saccharomyces cerevisiae

3. Sexual Sporulation Study (*S. cerevisiae*). Drawings of asci and ascospores (see Procedure in Second Session step 3).

4. Chlamydospore Formation Study (*C. albicans*). Drawings of chlamydospores (see Procedure in Second Session step 4).

5. Germ Tube Formation Study (*C. albicans*). Drawings of germ tubes (see Procedure in First Session step 5).

Questions

1. List four ways of differentiating *Candida albicans* from *Saccharomyces cerevisiae*.

2. What are two ways in which you were able to differentiate pathogenic from nonpathogenic *Candida* species?

3. What criteria would you use to differentiate the fermentation of glucose from the assimilation of glucose?

4. Why is a loop rather than a pipet used to inoculate the sugar fermentation tubes?

5. Why would the growth of a pellicle or film on the surface of a broth growth medium be advantageous to the physiology of that yeast?

EXERCISE

20

Filamentous Fungi: Particularly Microscopic Forms Such as Molds

Figure 20.1 The two major hyphal types found in fungi.

Nonseptate (coenocytic) hyphae

Septate hyphae

Figure 20.2 Asexual reproductive structure of the phycomycete *Rhizopus nigricans*. Note coenocytic hyphae.

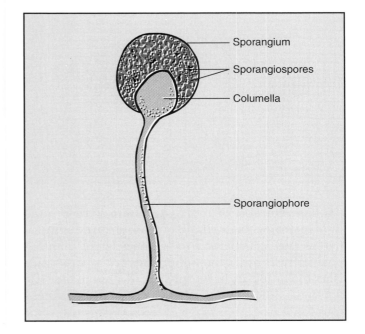

- Sporangium
- Sporangiospores
- Columella
- Sporangiophore

Getting Started

Filamentous fungi come in many sizes, shapes, and forms ranging from large bracket fungi found on trees, mushrooms, and puffballs to the small colonies often found as contaminants on moldy fruits and cheese (see color plates 8 and 9). Their diverse morphology, which can be observed both visually and with the aid of the microscope, makes for ease in their identification. Most fungi can be identified to at least the genus level, solely by studying their morphology. Physiological tests, which are very important in bacterial identification, are minimal if any in the study of filamentous fungi.

Structurally, the filaments or threads described as **hyphae** either contain septations with one or more nuclei per unit (multinucleate), or do not contain septations but are still considered multinu-

cleate **(coenocytic)** (figure 20.1). This structural difference is also important taxonomically since one of the four classes of fungi is distinguished from the other classes by its lack of septate filaments—the Class Phycomycetes, which contains only a few human pathogens.

A mass of filaments forms a mat described as **mycelium,** which can be seen with the naked eye when the fungus forms a colony on a solid medium. The hyphae submerged in the medium serve as an anchor for a later growth stage when the colorless hyphae form aerial fruiting structures with asexual and/or sexual spores, which are often colored. These fruiting structures contain diagnostic features important in making fungus identification. Asexual reproductive structures of three commonly found **molds** are shown in figures 20.2, 20.3, and 20.4.

Figure 20.3 Asexual reproductive structure of the ascomycete *Aspergillus niger*. Note the presence of a foot cell and a columella. Also note septate hyphae.

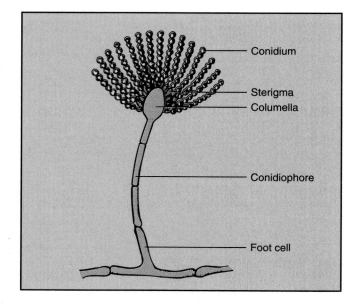

- Conidium
- Sterigma
- Columella
- Conidiophore
- Foot cell

Figure 20.4 Asexual reproductive structure of the ascomycete *Penicillium*. Note the absence of a columella and a foot cell. Also note the symmetrical attachment of the metulae to the conidiophore—an important diagnostic feature for differentiation within the genus. Also note septate hyphae.

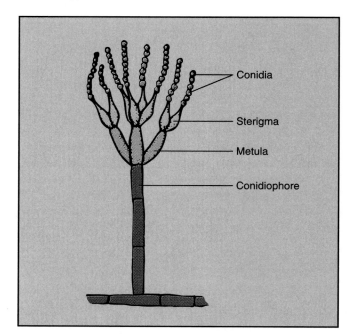

- Conidia
- Sterigma
- Metula
- Conidiophore

Examination of mold mycelium with a hand lens reveals that a single strand of mycelium, or hypha, has a diameter 5 to 10 times larger than hyphae formed by a group of prokaryotic microorganisms classified as actinomycetes. This diagnostic difference allows us to readily distinguish filamentous fungi from this group of microorganisms, some of which are pathogenic to humans (members of the genera *Nocardia* and *Actinomyces*). Since actinomycetes are filamentous, they are included as part of the study of clinical **mycology.**

The four classes* of fungi are shown in table 20.1.

Most of the medically important fungi are found in the Classes Ascomycetes and Deuteromycetes (fungi lacking a known sexual reproductive stage). Most of these infections are limited to cutaneous or subcutaneous tissues. However, other opportunistic infections can become progressive, leading to a systemic involvement and the possibility of death. According to Al-Doory, it is

*In recent years, the fungi have been divided into five rather than four classes. This change occurred because two sexual spore forms occur in the class Phycomycetes—the oospores and zygospores. The former class name was replaced by the new class names Oomycetes and Zygomycetes. Since that time, however, even further changes have occurred (see your text chapter 12, p. 282).

expected that the use of new medical technologies involving instrumentation, surgical procedures, and new drugs will continue, thus increasing the ever-present risk of opportunist fungal infection. This, in turn, will expand the need for trained mycologists and clinical mycology laboratories.

Clinically, there are three types of mycoses:

1. **Dermatomycoses** are superficial **keratinized** infections of the skin, hair, and nails caused by a group of filamentous fungi commonly called dermatophytes. They rarely invade subcutaneous tissues. **Parasitically,** they show rudimentary morphology, appearing only as mycelial growth in skin and nails, or as fragments of mycelium and **arthrospores** arranged inside and outside of hair. In all instances, they form circular lesions described as **ringworm** (see color plate 10).

   However, in culture they form filamentous colonies and asexual reproductive spores.

2. **Subcutaneous mycoses** are caused by either filamentous or by **dimorphic** yeastlike fungi.

Table 20.1   Classification of the Fungi

	Class			
	Phycomycetes	Ascomycetes	Basidiomycetes	Deuteromycetes
Mycelium	Nonseptate	Septate	Septate	Septate
Sexual spores	Oospore (not in a fruiting body) found in aquatic forms; Zygospore (not in a fruiting body) found in terrestrial forms	Ascospores, borne in an ascus, usually contained in a fruiting body	Basidiospores, borne on the outside of a clublike cell (the basidium), often in a fruiting body	None
Asexual spores	Zoospores, motile; Sporangiospores, nonmotile, contained in a sporangium	Conidiospores, nonmotile, formed on the tip of a specialized filament, the conidiophore	Same as Ascomycetes	Same as Ascomycetes
Common representatives	Downy mildews, potato blight, fly fungi, bread mold (*Rhizopus*)	Yeasts, morels, cup fungi, Dutch elm disease, ergot	Mushrooms, puff balls, toadstools, rusts, smuts, stinkhorns	Mostly imperfect Ascomycetes and some imperfect Basidiomycetes*

*Some of these fungi will no doubt form sexual spores in the right environment. In this event, they would be reclassified.

An example of such a fungus is *Sporotrichum schenkii*, which produces a fixed cutaneous sporotrichosis that remains restricted (fixed) to the site of infection (see color plate 11). Morphologically, it assumes a filamentous form when incubated at 20° C and a yeastlike form when incubated at 37° C (see color plates 12, 13, and 14).

3. **Systemic yeast** and **yeastlike infections** originate in the lungs, and can spread to other parts of the body. Examples include **histoplasmosis,** which is caused by the dimorphic yeastlike fungus *Histoplasma capsulatum,* and **coccidioidomycosis** (San Joaquin fever), which is caused by *Coccidioides immitis,* an organism exhibiting a form of dimorphism in which mycelium with **arthrospores** are produced in one phase of *in vitro* growth, and **culture spherules** containing asexual endospores (see color plates 15, 16, and 17) are produced when the CO_2 tension is increased in the growth medium.

Some Clinical Laboratory Diagnosis Procedures

Fungi responsible for mycotic diseases may be isolated from a variety of specimens including skin scrapings, sputum, spinal fluid, and lung tissue biopsy samples. It is important that the specimen is not contaminated with other microorganisms during the collection process; for example, in the collection of a sputum sample, the patient's teeth must be brushed thoroughly or his dentures removed, and the mouth cleansed by intensive use of a mouthwash. Only a specimen from deep in the lungs should be collected. Nasopharyngeal secretions, saliva, and other superficial oral specimens should not be collected. The specimen should be collected in a sterile container with a wide mouth to permit deposit of the sputum without undue difficulty. The first sputum specimen expectorated in the morning is usually the best, with a 10- to 15-ml sample being adequate.

Following culturing of material from the specimen, it is common to make a direct microscopic examination of it since this procedure often saves time in the diagnosis of a fungal disease. This usually consists of a wet mount preparation, which may or may not be stained. If the specimen is viscid or has an opaque appearance, it can first be digested with 10% KOH or an enzyme preparation.

Specimens to be cultured should be processed immediately. In any event, culturing should not be delayed more than 2 hours after collection. Long periods between collection and processing may lead to excessive overgrowth of rapidly growing organisms such as bacteria. The medium most commonly used for initial isolation of fungi is Sabouraud's dextrose agar with or without antibiotics to inhibit the growth of contaminant bacteria. The temperature of incubation depends on the organisms sought (*Aspergillus fumigatus*, for instance, grows well at 45° C, a temperature that inhibits growth of most other fungi) and composition of the medium (media containing the antibiotic cycloheximide should be incubated at 25° to 30° C since many pathogenic fungi are inhibited by cycloheximide at 37° C). Slide cultures are commonly used for visualizing various asexual and sexual reproductive structures since the growth on Petri dish cultures is usually too dense.

In addition, fungus diseases can be diagnosed by immunological methods (see exercise 32).

To provide you with an opportunity to identify filamentous fungi by their morphology, you will receive for study Petri dish cultures of the three nonpathogenic molds shown in figures 20.2, 20.3, and 20.4. Perhaps some previously prepared covered slide cultures will also be made available for your use in making a more precise study of their asexual fruiting structures.

Petri dish cultures usually do not afford such an opportunity since the growth is often too dense to see individual, intact fruiting structures. Also, when you attempt to prepare a wet mount from a Petri dish culture, the intact fruiting structures are usually broken apart, leaving only individual parts of the fruiting structure. These parts can often be helpful in identification, but are usually not sufficient to make clear-cut identifications. The Petri dish cultures serve best for study of colonial form as well as for wet mount observation of individual organ morphology such as spores.

Definitions

Arthrospores. Thick-walled asexual spores formed by breaking apart of the cells in septate hyphae.

Chlamydospore. A resistant hyphal cell with a thick wall; it eventually separates from the hypha and functions as a spore.

Coccidioidomycosis. An upper respiratory tract infection caused by the dimorphic yeastlike organism *Coccidioides immitis*.

Coenocytic. A multinucleate mass of protoplasm resulting from repeated nuclear division unaccompanied by cell division.

Conidia. Asexual spores produced at the tip or side of the conidiophore, or hypha.

Dermatomycosis. A disease of the skin caused by infection with a fungus.

Dimorphic. Capable of having two forms, as in some of the fungi that exist as either a mycelial or yeastlike form.

Foot cell. A cell located at the base of the conidiophore in the genus *Aspergillus*.

Histoplasmosis. A pulmonary infection caused by the dimorphic yeastlike organism *Histoplasma capsulatum*.

Hypha (pl. **hyphae**). Threadlike fungal filament(s) that form a mat described as mycelium.

Keratinized. Conversion of tissue into keratin or keratinized tissue. Forms the chemical basis of epidermal structures such as hair and nails that are insoluble in most solvents.

Mold. A multicellular filamentous fungus often appearing as a woolly growth on decaying materials.

Mycelium. A fungal mat made of tangled hyphae.

Opportunist. An organism capable of causing disease only when host defense mechanisms are impaired.

Parasite. An organism that lives on another organism (the host) and gains at the host's expense.

Pyriform. Pear shaped.

Rhizoids. Fungus hyphae that penetrate various substrates in order to obtain nutrients.

Ringworm. Any of several contagious fungal diseases of the hair, skin, or nails characterized on the skin by ring-shaped, discolored patches covered with vesicles and scales.

Spherules. A large, round, thick-walled structure filled with fungal endospores; characteristic of the tissue phase of *Coccidioides immitis*.

Sporangiospore. An asexual spore produced in the sporangium of certain Phycomycetes.

Sterigma (pl. **sterigmata**). A specialized hypha that supports either a conidiospore(s) or a basidiospore(s).

Objectives

1. To introduce you to the world of filamentous fungi and the methods commonly used for their identification. Included is a classification outline describing the four major classes of fungi and another outline describing the three major groups of medically important fungi.
2. To introduce some of the morphological methods commonly used for identification of filamentous fungi.

References

Al-Doory, Y. *Laboratory medical mycology*. Philadelphia: Lea and Febiger, 1980.

Nester et al. *Microbiology: A human perspective*. Chapter 12, 278–286.

Rippon, J. W. *Medical mycology*, 3d ed. Philadelphia: W. B. Saunders Co., 1988.

Materials

Cultures (to be shared by three students)

Petri dish cultures of *Rhizopus nigricans*, *Aspergillus niger*, and *Penicillium* species or *notatum* on Sabouraud's dextrose agar. For preparation, see Notes to Instructor.

Covered slide cultures of the above three fungi on Sabouraud's dextrose agar. Optional: see Notes to Instructor for preparation.

Dissecting microscope(s)

A ruler divided in mm

Procedure

Safety Precaution: When not working with the Petri dish cultures, keep them covered. Otherwise, excessive contamination throughout the room may occur since the spores are easily disrupted.

Colonial Characteristics of Petri Dish Cultures

1. Visually examine each Petri dish culture noting the following:
   a. Colony size. With a ruler measure the diameter in mm.
   b. Colony color. Examine both the upper and lower surfaces.
   c. Presence of soluble pigments in the agar medium.
   d. Colony texture (such as cottony, powdery, or woolly).
   e. Colony edge (margin). Is it regular or irregular?
   f. Colony convolutions (ridges). Are they present?

Insert your findings in table 20.2 of the Laboratory Report.

Morphology of Asexual Reproductive Structures

Note: A dissecting microscope is preferable for use in making your initial observations. If one is not available, the light microscope can be used. If so, remove the cover with care and place the uncovered dish on the stage of the microscope, where it can be observed initially with your lowest power objective. The cover should be placed on a piece of disposable toweling because the fruiting bodies sometimes become attached to the cover surface, particularly with the genus *Rhizopus*. If this occurs, microscopic examination of the inverted cover is another possibility for study of intact fruiting structures.

1. Place the *covered* Petri dish culture on the stage of the dissecting microscope and examine a sparsely populated area of a colony for the presence of asexual reproductive structures (see figures 20.3 and 20.4).
2. When you first observe fruiting bodies, STOP.

   Note: Never smell fungal cultures—spore inhalation can cause infection.

   Keeping air currents to a minimum to avoid spore dispersal, carefully remove the Petri dish cover and re-examine them to determine whether you can in fact see the various parts of the fruiting body as described in the figure for that fungus. If not, you may find it necessary to use a covered slide culture. You may also wish to examine the Petri dish culture with the low power objective of the light microscope, which more than doubles the magnification obtained with the dissecting microscope.

Note: Use of the high power objective inside the Petri dish will prove difficult. However, it can be used if you need to examine the inside of a Petri dish cover. Be careful not to touch the fungus with the high power objective. If this occurs, first focus with the low power objective, then and only if parfocal, switch to the high power objective.

3. Prepare drawings of the intact asexual reproductive structures in part 2 of the Laboratory Report. Label the various parts as shown in the figures.

4. Detailed examination of **sporangiospores, conidia,** and, if present, **chlamydospores:** For observing sporangiospores (*Rhizopus*) and conidia (*Aspergillus* and *Penicillium*), remove some aerial growth with a loop. Then prepare a wet mount and observe it with the low and high dry objectives of the microscope. Chlamydospores can be found in the *Rhizopus* surface and submerged mycelium. They are elongated, brown in color, and have thick walls. Prepare a wet mount and observe with the low and high dry objectives.

Note: You may wish to first search for them in both the inner and outer fringes of the colony using the low power objective. To search the inner colony surface, make the area less dense by first removing some of the aerial growth with a loop. Flame the loop to destroy the spores. Prepare drawings of the various spore types found in part 3 of the Laboratory Report.

Some morphological characteristics of value in identification are:

1. *Rhizopus nigricans*
   a. Has nonseptate coenocytic hyphae
   b. Contains **rhizoids,** which are similar in function to stolons found in strawberries. Look for them on the underside of the Petri dish of an older culture.
   c. Details of fruiting body (figure 20.2). Note the nonseptate stem (sporangiophore), the swelling at the tip of the sporangiophore **(columella),** and the sac that encloses the columella **(sporangium)** and contains the asexual reproductive spores (sporangiospores).

2. and 3. *Aspergillus niger* and *Penicillium notatum*
   These molds have fruiting bodies somewhat similar in appearance. Both have brushlike structures comprised of bottle-shaped cells **(sterigmata)** to which are attached long chains of the asexual reproductive spores **(conidiospores).** They differ in that the genus *Aspergillus* has a swollen cell at the base of the stem **(conidiophore)** known as the **foot cell,** as well as a columella to which the sterigmata are attached (see figure 20.3). The sterigmata may occur in one or two series depending upon the species involved. Finally, in the genus *Penicillium* the branching at the tip of the conidiophore can be either **symmetrical** or **asymmetrical** (see figure 20.4), depending on the point of attachment of the metulae to the conidiophore. When all of the metulae are attached at the tip of the conidiophore, branching of the sterigmata and attached conidia will appear symmetrical (see figure 20.4). If one of the 3 metulae in figure 20.4 is attached below the tip of the conidiophore, an asymmetrical branching occurs. This is an important diagnostic feature for differentiation within the genus *Penicillium.*

Name _____ Date _____ Section _____

Laboratory Report: Filamentous Fungi:
Particularly Microscopic Forms Such as Molds

Results

1. Colonial characteristics of *Rhizopus*, *Aspergillus*, and *Penicillium* when grown on Sabouraud's dextrose agar.

Table 20.2   Colonial Characteristics of Three Filamentous Fungi

	Rhizopus	*Aspergillus*	*Penicillium*
Colony color			
Colony diameter (mm)			
Colony texture			
Colony convolutions			
Colony margin			
Soluble pigments in agar			

2. Drawings of their asexual reproductive structures.

    Rhizopus   *Aspergillus*   *Penicillium*

3. Drawings of their asexual spores.

    Rhizopus   *Aspergillus*   *Penicillium*

Questions

1. What are some ways in which you might be able to differentiate *Rhizopus nigricans* from *Aspergillus niger* simply by visually observing a Petri dish culture?

2. How could you determine whether or not a green, woolly looking colony was an *Aspergillus* or *Penicillium?*

3. What problem might you have in identifying a pathogenic fungus observed in a blood specimen? What might you do to correct the problem?

4. In what ways can we readily distinguish:
   a. fungi from algae?

   b. fungi from bacteria?

   c. fungi from actinomycetes?

5. Provide an example of an opportunistic fungus. Are all medically important fungi opportunistic?

6. Name three pathogenic fungi that exhibit dimorphism. Describe the type of dimorphism each exhibits and the laboratory conditions necessary to elicit it.

Sabauraud agar — on slide — lay coverslip over
grow fungi at room Temp ☐ *chunk tap*

Fungi — Mycology
4 Basic classes — classified by mode of sexual reproduction
1) Phycomycetes — naked spores
2) Ascomycetes — ascospores — produced in sac-like asms.
3) Basidiomycetes — basidiospore
4) Deuteromycetes — no known sexual reproduction

septate   coenocytic — not septate

mycelium : contain hyphae
aerial — grow upwards
vegatative — stay down near mycelium nut — supportive layer

1) superficial — grow outside
2) cutaneous — able to produce keritanase
   dermatophytes — ringworm
      Tinea capitis,   corporis — body
         ↳   pedis → feet
         body part   ungium — nails
      ↳ ringworm (fungus)

3) subcutaneous — deep layers of skin
   soil saprophytes
4) Systemic — true pathogens
   dimorphic fungi — 2 growing characteristics
   Room temp (25°C) — mold
   Body temp (37°C) — yeast phase
   inhalation of spore
   fungi — glass petri dishes

5) Opportunistic — Candida albicans
   Penicillium / Aspergillus / Rhizopus / Alternaria

EXERCISE 21

Parasitology: Protozoa and Helminths

Getting Started

Since the natural histories of parasitic diseases differ in some important respects from those of bacterial diseases, they merit a separate laboratory session to give you at least minimal direct experience with parasites, the diseases they cause, and techniques used to diagnose them.

The distinguishing feature of parasitic life is the close contact of the parasite with the host in or on which it lives, and its dependency on the host for life itself. This special association has led to the evolution of three types of adaptations not found in the free-living relatives of the parasites: loss of competency, special structures, and ecological ingenuity.

Parasites have become so dependent on their hosts for food and habitat that they have experienced a *loss of competency* to live independently. They usually require a specific host and many have lost their sensory and digestive functions; these are no longer important for their survival.

On the other hand, they have developed *special structures* and functions not possessed by their free-living relatives, which promote survival within the host. One example is special organs of attachment—hooklets and suckers. Parasites also have a tremendously increased reproductive capacity, which compensates for the uncertainty in finding a new host. Tapeworms, for example, have fantastically high rates of egg production, reaching up to 100,000 per day.

Ecological ingenuity is demonstrated in the fascinating variety of infecting and transmitting mechanisms. This has led to very complex life cycles, which contrast markedly with the relatively simple life styles of their free-living counterparts. Parasites show quite a range in the types of life cycles they possess, from species that pass part of each generation in the free-living state to others that require at least three different hosts to complete the life cycle. Some are simply transmitted by insects from one human host to the new host, or the insect may act as a host as well. Many protozoa develop resistant cysts that enable them to survive in unfavorable environments until they find a new host. The eggs of flatworms and roundworms also have a protective coat.

These three strategies promote survival and expansion of the species by providing greater opportunities for finding and infecting new hosts, which is a continual problem for parasites. Successful interruption of these cycles to prevent their completion is an important feature of public health measures used to control diseases caused by parasites.

This exercise is designed to give you some practical experience with representative protozoan and helminthic parasites, and with clinical methods used in their diagnosis and control. Your study will consist of these procedures:

1. As an introduction, you will have an opportunity to observe the movements and structure of some living nonparasitic protozoans and worms often found in pond water.

2. Examination of commercially prepared stained blood and fecal slides that contain human protozoan parasites.

3. Microscopic comparison of the structure of parasitic worms with that of their free-living relatives to observe some special adaptations to the parasitic way of life.

4. Study of the natural history and life cycle of the human parasitic disease schistosomiasis. This will enable you to see the interaction between stages of the life cycle, environmental milieu, and social conditions of their human hosts as factors in the epidemiology and control of the disease.

The following simplified classification of parasites will serve as a guide to the examples you will be studying in this exercise. It is not a complete listing.

Protozoa

Protozoa, a subkingdom of the kingdom Protista, are unicellular eukaryotic organisms. They usually reproduce by cell division and are classified into five phyla mainly according to their means of locomotion. Only one of these phyla, the Phylum Suctoria, which is closely related to the Phylum Ciliata, does not contain animal pathogens. The remaining phyla are classified as follows:

Phylum Sarcodina

Members of this phylum move and feed slowly by forming cytoplasmic projections known as **pseudopodia** (false feet). They also form both **trophozoites** (vegetative form) and **cysts** (resistant, resting cells). Parasitic members include the amoeba *Entamoeba histolytica*, which causes amoebic dysentery. It ingests red blood cells and forms a four-nucleate cyst. It is also found in animals. Other amoeba species found in humans, such as *Entamoeba gingivalis*, are relatively harmless **commensals.**

Phylum Ciliata

Members of this phylum have many short, hairlike cilia on their body surfaces that beat rhythmically by bending to one side. They contain two nuclei: a **macronucleus** and a **micronucleus.** This phylum is typified by the genus *Paramecium* (see color plate 18), a nonpathogenic form readily found in pond water. Another member, *Balantidium coli*, is a common parasite in swine, and can infect humans, causing serious results.

Phylum Mastigophora

These protozoans propel themselves with one or more long, whiplike flagella. Some have more than one nucleus and usually produce cysts. Different species cause infections in the intestines, vagina, blood, and tissues. *Giardia lamblia* (see figure 34.4) causes a mild to severe diarrheal infection. *Trichomonas vaginalis* (figure 21.1) is found in the urogenital region, where it causes a mild vaginitis in women. *Trypanosoma gambiense* (figure 21.2) infects the blood via tsetse fly bites, where it causes trypanosomiasis, or African sleeping sickness, in cattle and humans. Cattle and other ungulates serve as a reservoir for this organism.

Figure 21.1 *Trichomonas vaginalis*. Illustration of a typical mastigophoran protozoan. From Eugene W. Nester, C. Evans Roberts, and Martha T. Nester, *Microbiology: A Human Perspective.* Copyright © 1995 Wm. C. Brown Communications, Inc., Dubuque, Iowa. All Rights Reserved. Reprinted by permission.

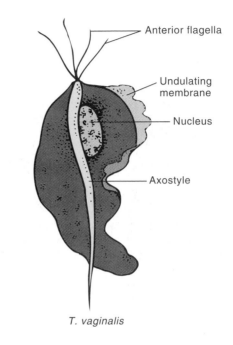

T. vaginalis

Figure 21.2 A prepared blood smear of trypanosome parasites in human blood. The slender flagellates lie between the red blood cells. © Cleveland P. Hickman, Jr./Visuals Unlimited

Phylum Sporozoa

These are obligate, nonmotile parasites with alternating stages: the sexual reproductive stage is passed in the *definitive* insect host and the asexual phase in the *intermediate* human or animal host. The genus *Plasmodium* includes the malarial species, in which

Figure 21.3 The genus *Dugesia,* a free-living planarian in the class Turbellaria. (*a*) A living specimen (✕ 0.5) is shown, and (*b*) shows a labeled line drawing. (*a*) © John D. Cunningham/Visuals Unlimited

(a)

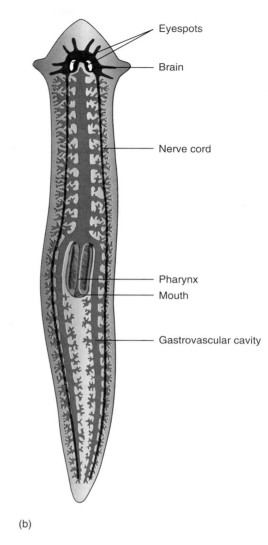

- Eyespots
- Brain
- Nerve cord
- Pharynx
- Mouth
- Gastrovascular cavity

(b)

the definitive host is the female *Anopheles* mosquito, and the intermediate host humans. The genus *Coccidia* includes important intestinal parasites of fowl, cats, dogs, swine, sheep, and cattle. *Toxoplasma* species is a cat parasite that can harm the human fetus in an infected pregnant woman.

Helminths (Worms)

These are multicellular eukaryotic organisms. Two of the phyla, Platyhelminthes (flatworms) and Nemathelminthes (roundworms) contain pathogenic worms.

Members of the Platyhelminthes are flat, elongated, legless worms that are **acoelomate** (without a body cavity), and exhibit bilateral symmetry. This phylum contains three classes:

Class Turbellaria

These are free-living **planarians** (flatworms), such as are found in the genus *Dugesia* (figure 21.3).

Class Trematoda (Flukes)

They have an unsegmented body and many have suckers to hold them onto the host's intestinal wall. Many flukes have complex life cycles that require aquatic animal hosts. The *Schistosoma* species are bisexual trematodes that cause serious human disease.

They require polluted water, snails, and contact with human skin for completion of their life cycles (see color plates 19 and 20). *Clonorchis sinensis* and *Fasciola* species are liver flukes acquired by eating infected raw fish and contaminated vegetables, respectively.

Class Cestoda (Tapeworms)

These are long, segmented worms with a small head **(scolex)** equipped with suckers and often hooklets (figure 21.4) for attachment to the host's intestinal wall. The series of segments, or **proglottids,** contain the reproductive organs and thousands of eggs. These segments break off and are eliminated in the

Figure 21.4 Illustration of a tapeworm scolex showing both hooklets and suckers for attachment to the intestine. *T. saginata* (beef tapeworm) is essentially without hooklets, whereas *T. solium* (pork tapeworm) has both.
© Stanley Flegler/Visuals Unlimited

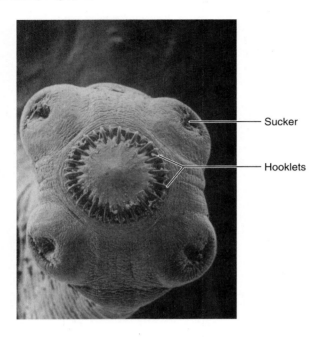

feces, leaving the attached scolex to produce more proglottids with more eggs. Figure 21.5 illustrates the life cycle of the tapeworm in a human. The symptoms of *Taenia* tapeworm are usually not serious, causing only mild intestinal symptoms and loss of nutrition. Not so for the *Echinococcus* tapeworm, which causes a serious disease. All tapeworm diseases are transmitted by animals.

Members of the Phylum Nemathelminthes (roundworms) occupy an important ecological niche since they are present in large numbers in very diverse environments, including soil, fresh water, and seawater. In contrast to the Platyhelminthes, these round, unsegmented worms are **coelomate** (have a body cavity), and have a complete digestive tract and separate sexes. This phylum contains many agents of animal, plant, and human parasitic diseases. Most require only one host, and can pass part of their life cycle as free-living larvae in the soil. *Trichinella spiralis* requires alternate vertebrate hosts. Humans become infected when they ingest inadequately cooked meat, such as pork or wild game, containing the larval forms in the muscles. *Ascaris lumbricoides*

Figure 21.5 Life cycle of *Taenia saginata*. The adult tapeworm with scolex and proglottids is conceived from larvae in the human intestine.

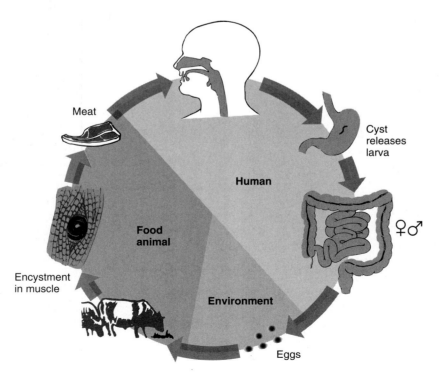

21–4 Exercise 21 Parasitology: Protozoa and Helminths

Figure 21.6 *Ascaris lumbricoides,* an intestinal roundworm up to 12 inches long. A mass of worms recovered from the ileum of a malnourished child. From Rubin and Farbo, *Pathology.* Reprinted by Permission of J. B. Lippincott Company.

(figure 21.6) is probably the most common worldwide of all the human helminths. *Enterobius vermicularis* causes pinworm, a very common condition in children in the United States. Efforts to eradicate it have not been very successful since pinworm causes little discomfort. *Oxyuris* causes a similar condition in animals.

Definitions

Acoelomate. Without a true body cavity. Typical of members of the Phylum Platyhelminthes (flatworms).

Amoeba. Unicellular organisms with an indefinite changeable form.

Cercaria. The last miracidium stage in which the larvae possess a tail.

Coelomate. With a true body cavity. Typical of members of the Phylum Nemathelminthes (roundworms).

Cysts. Dormant, thick-walled vegetative cells.

Definitive host. The host in which the sexual reproduction of a parasite takes place.

Intermediate host. The host that is normally used by a parasite in the course of its life cycle, and in which it multiplies asexually but not sexually.

Merozoites. Schizont nuclei that become surrounded by cytoplasm and bud off as daughter cells or merozoites.

Miracidium. A free-swimming ciliate larva that seeks out and penetrates a suitable intermediate snail host in which it develops into a sporocyst.

Planarian. Any flatworm of the genus *Planaria.*

Proglottids. Any of the segments of a tapeworm formed in the neck region by a process of strobilation (transverse fission).

Pseudopodia. Extensions of cytoplasm that aid in engulfing particles and functioning in motility of amoeboid cells.

Schizont. A stage in the life cycle of Sporozoa in which the nucleus of the parent cell, the schizont, undergoes repeated nuclear division without corresponding cell divisions.

Scolex. The head of a tapeworm, which is used for attaching to the host's intestinal wall.

Sporocyst. A stage in the life cycle of certain protozoa in which two or more of the parasites are enclosed within a common wall.

Trophozoites. Vegetative forms of some protozoans.

Objectives

1. To introduce you to the study of parasitology by letting you examine some examples of nonparasitic protozoans (*Amoeba proteus* and a *Paramecium* species) and a free-living planarian (flatworm) of the genus *Dugesia.*

2. To study the morphology of some free-living, trophozoite, and cystic forms of intestinal

Figure 21.7 *P. falciparum* infection showing the ring trophozoite stage in a blood smear. Centers for Disease Control

Ring trophozoite

parasites using prepared slides, and malarial and trypanosome parasites using stained blood smears.

3. To study the natural history and life cycle of an important human parasitic disease, schistosomiasis, using stained slides and a life cycle diagram (figure 21.7).
4. To demonstrate special adaptations of parasitic worms through a study of stained slides in which they are compared to similar free-living relatives.

References

Leventhal, R., Cheadle, R. *Medical parasitology*, 3d ed., F. A. Davis, Philadelphia, 1989.

Markell, E., Voge, M., John, D. *Medical parasitology*, 7th ed., Saunders, Philadelphia, 1992.

Nester et al. *Microbiology: A human perspective*. See Appendix IX Arthropods and Parasitic Worms.

Neva, F. A., Brown, H. W. *Basic Clinical Parasitology*, 6th ed., Appleton Lange, Norwalk, 1994.

Materials

Cultures

Living cultures of a *Paramecium* species, *Amoeba proteus*, and a *Dugesia* species

If available, a fresh sample of quiescent, stagnant pond water, which often contains members of the above genera.

Planaslo solution, 1 or more dropping bottles

Depression slides (hanging drop slides)

The following commercially prepared slides:

Subkingdom Protozoa

Phylum Sarcodina (pseudopodia)
Entamoeba histolytica trophozoite and cyst stages
Phylum Ciliata (cilia)
Paramecium trophozoite
Phylum Mastigophora (flagella)
Giardia lamblia trophozoite and cyst stages
Trypanosoma gambiense
Phylum Sporozoa (nonmotile)
Plasmodium vivax ring, amoeboid schizont stages

Subkingdom Helminths (worms)

Phylum Platyhelminthes (flatworms)
Class Turbellaria (free living)
Dugesia species
Class Trematoda (flukes)
Schistosoma mansoni
  adult male
  adult female
  ovum (egg)
  ciliated miracidium
  infective ciliate cercaria
  sporocyst stage in snail liver tissue
Clonorchis sinensis
Class Cestoda (tapeworms)
Taenia solium trophozoite

Procedure

Note: If the number of prepared slides is limited, these procedures may be performed in a different order to facilitate sharing of slides.

1. Examination of free-living cultures
   a. Pond water examination. Prepare a hanging drop slide (see exercise 3). Examine initially with the low power objective and later with the high and oil-immersion objectives. Observe the mode of locomotion of any amoeboid or paramecium-like protozoans found. If their movements are too rapid, add a drop of Planaslo to slow them down. Describe their movements and prepare drawings in part 1a of the Laboratory Report.
   b. Examination of fresh samples of an amoeba (such as *Amoeba proteus*), paramecium (for example, *Paramecium caudatum*), and a free-living flatworm (such as *Dugesia* species). Use hanging drop slide preparations and examine as described above for pond water. Record your observations in part 1b of the Laboratory Report.

Note: Although less desirable for observing motion, wet mounts may be substituted if necessary for hanging drop slides.

2. Examination of stained slides for trophozoites and cysts
   a. Using the oil-immersion objective, examine prepared slides of a protozoan, either the amoeba *Entamoeba histolytica* or the flagellate *Giardia lamblia*. In the trophozoite stage, observe the size, shape, number of nuclei, and presence of flagella or pseudopodia. In the cyst stage, look for an increased number of nuclei and the thickened cyst wall.
   b. Sketch an example of each stage, label, and record in part 2 of the Laboratory Report.

3. Examination of protozoans present in stained blood slides
   a. Examine with oil immersion a smear of blood infected with *Plasmodium vivax* and locate blood cells containing the parasite. After a mosquito bite, the parasites are carried to the liver, where they develop into merozoites. Later, they penetrate into the blood and invade the red blood cells, where they go through several stages of development. The stages are the delicate ring stage (see figure 21.7), the mature amoeboid form, and the schizont stage, in which the organism has divided into many individual infective segments that will then cause the red cell to rupture, releasing the parasites, which can then infect other cells. Sketch the red cells with the infective organism inside them, and note any changes in the red cell shape, pigmentation, or granules due to the effect of the parasite. Identify and label the stage or stages seen, and the species in part 3a of the Laboratory Report.
   b. Examine the trypanosome blood smear (see figure 21.2) with the oil-immersion lens, and locate the slender flagellates between the red blood cells, noting the flagellum and undulating membrane. Sketch a few red cells along with a flagellate in part 3b of the Laboratory Report.

4. Comparison of a free-living worm with its parasitic relative
   a. Observe a prepared slide of a free-living flatworm (*Dugesia* species) with the low power objective. Note the pharynx, digestive system, sensory lobes in the head region, and the eyes (see figure 21.3). Next examine a parasitic fluke such as *Clonorchis*. Note the internal structure, especially the reproductive system and eggs if female, and the organs of attachment such as hooklets or round suckers.
   b. Sketch each organism in part 4a of the Laboratory Report, and label the main features of each. Describe the main differences between the fluke and the free-living planaria.
   c. Examine prepared slides of a tapeworm (*Taenia* species, see figure 21.4), observing the small head, or scolex, and the

attachment organs—the hooklets or suckers. Then locate along the worm's length the maturing proglottids. The smaller proglottids may show the sex organs better; a fully developed proglottid shows the enlarged uterus filled with eggs. Sketch, label, and describe its special adaptations to parasitic life in part 4b of the Laboratory Report.

5. Life cyle of *Schistosoma mansoni* and its importance in the control of schistosomiasis

Assemble five or six slides showing the various stages in the schistosoma life cycle: adult worm (male and female if available), ova, ciliated miracidium, the sporocyst in the snail tissue, and the infective ciliate cercaria.

Next read this brief summary of the natural history of *Schistosoma mansoni* (see color plates 19 and 20 and Nester et al., Figure IX–20):

Human schistosomiasis occurs wherever these conditions exist: water is polluted with human wastes; this water is used for human bathing and wading, or irrigation of cropland; and the presence of snail species that are necessary as hosts for the sporocyst stage in fluke development and completion of its life cycle. Solution to this public health problem is very complex, not only because of technical difficulties in its control and treatment, but also because its life cycle presents an ecological dilemma. Many developing countries need food desperately, but the main sources now available for these expanding needs are fertile deserts, which have adequate nutrients but require vast irrigation schemes, such as the Aswan Dam in Egypt. However, due to the unsanitary conditions and the presence of suitable snail hosts, these projects are accompanied by an increase in the disease schistosomiasis, which currently is very difficult to control and very expensive to treat on a wide scale.

The *cercaria* larvae swim in the contaminated water, penetrating the skin of agricultural workers who are barefoot. They migrate into the blood and collect in the veins leading to the liver. The *adults* develop there, copulate, and release the *eggs*. The eggs are finally deposited in the small veins of the large intestine, where their spines cause damage to host blood vessels. Some eggs die; however, others escape the blood vessels into the intestine and pass with the feces into soil and water. There they develop, and then hatch into motile *miracidia*, which eventually infect suitable snail hosts and develop into saclike *sporocysts* in the snail tissues. From this stage develop the fork-tailed *cercaria* larvae, which leave the snail and swim in the water until they die or find a suitable human host, thus completing the complex life cycle involving two hosts and five separate stages.

Now look at the prepared slides of all the schistosoma stages discussed in the description above. Sketch each stage in the appropriate place in the life cycle diagram shown in part 5 of the Laboratory Report.

EXERCISE

Laboratory Report: Parasitology: Protozoa and Helminths

Results

1. Examination of Free-Living Cultures
   a. Pond water examination.
      Description of movements and drawings of any protozoans found in pond water.

   b. Examination of fresh samples of a free-living amoeba, paramecium, and flatworm. Description of movements and drawings with labels.

2. Examination of Stained Slides for Trophozoites and Cysts
   Prepare drawings of the trophozoite and cyst stage of either *Entamoeba histolytica* or *Giardia lamblia*. Label accordingly (see Procedure 2).

3. Examination of Protozoans Present in Stained Blood Slides
   a. Examine blood smears of *Plasmodium vivax* (see Procedure 3a).

   b. Blood smear of *Trypanosoma gambiense* (see Procedure 3b).

4. Comparison of a Free-Living Worm with Its Parasitic Relative
   a. Comparison of *Dugesia* species (free living) with *Clonorchis sinensis* (parasitic). See Procedure 4a.

   b. Study of a parasitic tapeworm (*Taenia* species). See Procedure 4b.

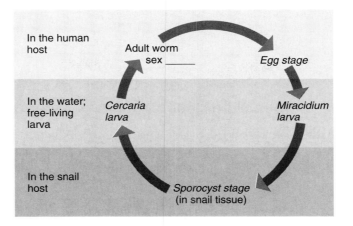

In the human host — Adult worm sex _____ — Egg stage

In the water; free-living larva — Cercaria larva — Miracidium larva

In the snail host — Sporocyst stage (in snail tissue)

5. Life Cycle of *Schistosoma mansoni* and Possible Methods of Control. In each space in this life cycle, sketch the appropriate stage, using the prepared microscope slide.

Proposed plan for public health control of schistosomiasis. Describe various strategies that might be developed by public health personnel to interrupt this cycle and thus prevent schistosomiasis. Show on the diagram where specific measures might be taken, and label. Explain each possibility and its advantages and disadvantages.

Questions

1. Which form—the trophozoite or the cyst—is most infective when found in a feces sample? Explain.

2. In what ways are free-living and parasitic worms similar, such that they can be identified as closely related?

3. In what ways do the parasitic species differ from the free-living planaria? Use the chart to summarize your comparisons.

	Planaria	Fluke	Tapeworm
Outside covering			
Organs of attachment			
Sensory organs			
Digestive system			
Reproduction			

4. Estimate the length and width of a trypanosome. See figure 21.2 for a clue. Show your calculations.

5. How is the *Echinococcus* tapeworm transmitted to humans? Does it cause a serious disease? What are two ways in which its transmission to humans can be prevented?

Figure 22.1 (*a–f*) Steps in the replication of a T-even phage during the infection of *E. coli.* From Eugene W. Nester, C. Evans Roberts, and Martha T. Nester, *Microbiology: A Human Perspective.* Copyright © 1995 Wm. C. Brown Communications, Inc., Dubuque, Iowa. All Rights Reserved. Reprinted by permission.

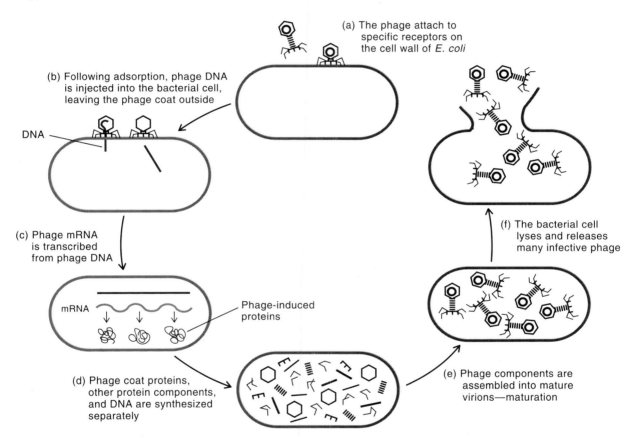

(a) The phage attach to specific receptors on the cell wall of *E. coli*

(b) Following adsorption, phage DNA is injected into the bacterial cell, leaving the phage coat outside

DNA

(c) Phage mRNA is transcribed from phage DNA

mRNA

Phage-induced proteins

(d) Phage coat proteins, other protein components, and DNA are synthesized separately

(e) Phage components are assembled into mature virions—maturation

(f) The bacterial cell lyses and releases many infective phage

Getting Started

Bacteriophage (usually shortened to phage) are viruses that infect bacteria. There are many kinds of bacteriophage, but this discussion is limited to DNA phage that have been well studied, such as lambda and T4. These phage first attach to the bacterial cell and inject their DNA into the cytoplasm. There are two major outcomes of this injection, depending on whether the phage are lytic or lysogenic.

1. **Lytic.** The cell lyses about 30 minutes after infection, releasing about 100 virus progeny (figure 22.1).

2. **Lysogenic** (or temperate). The DNA of the phage integrates into the bacterial chromosome and is replicated with the bacterial DNA. It may at some time in the future leave the chromosome, and direct the production of virus and lyse the cell. Bacteria that contain the DNA of a phage cannot be reinfected or lysed by the same type of phage.

Phage are too small (about 200 nm) to be seen in a light microscope, but can be detected if grown on a lawn of bacteria as follows. Phage and their host cells are mixed in a small tube of soft agar and then poured on top of an agar base plate. (Soft agar

Figure 22.2 Bacteriophage plaques formed on agar medium seeded with a lawn of bacteria. Courtesy of the University of Washington Photo Library

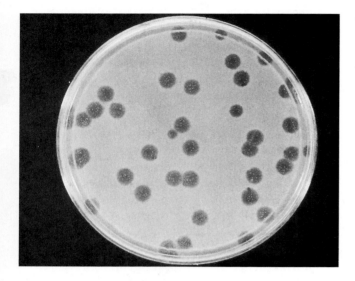

contains about half the concentration of standard agar so that the phage can diffuse more easily.) The plates are then incubated overnight at the optimum growth for the host bacteria.

During incubation, bacteria multiply and produce a thick covering of bacteria, or **lawn of bacteria,** except in those places where phage have infected and killed the bacteria leaving clear areas called **plaques** (figure 22.2). Since each plaque originated with one phage, the plaques can be counted just as bacterial colonies to determine the number of phage originally mixed with the soft agar. The number of phage per ml is referred to as the **titer.**

Although the appearance of the plaques can be influenced by many factors, in general **virulent phage** produce clear plaques. **Temperate phage** produce cloudy plaques because many cells within the plaque were lysogenized instead of lysed and thus continue to grow and multiply. The plaques do not increase in size indefinitely because phage can replicate only in multiplying bacteria. *Bdellovibrio,* a small ($1 \times 0.25 \mu$m) bacterium that multiplies within bacterial cells, also forms plaques. Since it can grow in nondividing cells, its plaques continue to increase in size for a day or two. It is also found in sewage and can pass through the filters used to remove bacteria, and may be present in the filtrate you use for phage isolation.

It is important to study phage and to learn the techniques used to manipulate them for the following reasons.

1. Animal viruses, including human pathogens, are grown on tissue culture cells in the same fashion as phage on bacteria. Tissue culture are animal cells that are grown in bottles and plates. The animal virus can form plaques by causing cells to degenerate or die. Tissue culture cells require a more complex (and more expensive) medium, so it is convenient to learn viral technique with phage and bacteria.
2. Phage are used in recombinant DNA experiments and are also useful in studying the genetics of bacteria.
3. They are used to identify different strains of bacteria because one type of phage will only infect a few specific strains.
4. Lysogeny has served as a model for viruses inserting their DNA in animal cells. The life cycle of temperate phages resembles animal retroviruses.

In this exercise, you will attempt to isolate phage that infect *Escherichia coli* and learn to titer phage. Since sewage contains high numbers of *E. coli,* it is an excellent source of the *E. coli* phage. The sewage is filtered to remove bacteria but not the smaller viruses. Samples of the filtrate are then mixed with a laboratory strain of *Escherichia coli* in a suitable medium and observed for plaques on agar plates. In the second part of the exercise, a suspension of phage is **serially diluted** so that an appropriate number of plaques can be counted on a plate and the titer of phage can be calculated.

Definitions

Bacterial lawn. The confluent growth of bacteria on an agar plate.

Bacteriophage. A virus that infects bacteria; often abbreviated phage (rhymes with rage).

Lysogenic phage. A phage that can establish a lysogenic relationship with bacteria by integrating into the bacterial chromosome. See temperate phage.

Lytic virus. A virus that replicates within a host cell and causes it to produce phage, rupture, and die. See virulent phage.

Plaque. A clear or cloudy area in a lawn of bacterial cells caused by phage infecting and lysing bacteria.

Plaque-forming units. The single phage that initiates the formation of a plaque.

Serial dilution. A dilution of a dilution continuing until the desired final concentration is reached.

Temperate, or lysogenic, phage. A phage that can either integrate into the host cell DNA or replicate outside the host chromosome, producing more phage and lysing the cell.

Titer. The concentration of a sample (number/volume) when used as a noun, or to determine the concentration when used as a verb.

Virulent, or lytic, phage. A phage that always causes lysis of the cell following phage replication. Unlike a temperate phage, it cannot be integrated into the chromosome of the host.

Objectives

1. To isolate a phage from its natural habitat.
2. To titer a phage suspension.
3. To have an appreciation of the phage life cycle (to view steps in a phage life cycle).

References

Fraenkel-Convar, H. *Virology.* Englewood Cliffs, N.J.: Prentice-Hall, 1982.
Maniatis et al. *Molecular cloning,* 2d ed. Volume I. Cold Spring Harbor Laboratory Press, 1989.
Nester et al. *Microbiology: A human perspective.* Chapter 13, 293–304.
Safety Precautions: Filtered sewage may contain harmful animal virus. Handle with extreme care.

Materials

First Session

Raw sewage filtered through 0.45 μm membrane filter

Host bacterial *Escherichia coli* K12 in a late log phase (OD~ 0.4)

Tubes soft tryptone overlay agar, 2

Tryptone agar base plates, 2

Part I: Bacteriophage Isolation and Culture from Raw Sewage

Procedure

First Session

1. Label plates 1 ml and 0.1 ml (figure 22.3).
2. Melt overlay agar in boiling water and place in 50° C water bath for at least 5 minutes.
3. Add 1ml filtered sewage to one tube soft overlay.
4. Add 0.1 ml filtered sewage to other tube of soft overlay.
5. Quickly add several drops (about 0.1 ml) of *E. coli* to each tube. Mix tubes and pour onto previously labeled tryptone agar base plates and rock gently to completely cover surface. If the tubes of overlay agar cool below 45° C, they will harden and the procedure must be repeated from the beginning.
6. Permit to harden for 5 minutes.
7. Incubate inverted at 37° C overnight.

Second Session

1. Examine the plates for evidence of plaque formation. Notice any different types of plaques and their relative sizes due to different kinds of phages. Count the number of each kind by placing dots with a marking pen under the plaques on the bottom of the Petri plates. Record results.

Figure 22.3 Schematic showing dilution procedure for isolating bacteriophage from enriched sewage.

Enriched sewage filtrate containing phage

E. coli culture

1.0 ml   0.1 ml   0.1 ml   0.1 ml   0.1 ml

Three tubes of soft agar in a 50°C water bath

Tubes of inoculated soft agar poured over plates of hard tryptone agar and incubated at 37°C

Plates may be observed after overnight incubation for the presence of viral plaques

E. coli control plate (one team)

2. Draw a circle the same size of each plaque on the bottom of the plate and reincubate. If any plaques are formed by *Bdellovibrio* the plaques will continue to enlarge. Phage plaques will remain the same size because, unlike *Bdellovibrio*, they can only reproduce in actively dividing cells.

Third Session

1. Examine plates for plaques that have increased in size.
2. If you do suspect *Bdellovibrio*, stab the plaque with a loop and prepare a wet mount. Look for very small, motile bacteria.
3. Record results.

Part II: Titering a Phage Suspension

Materials

Host bacteria for phage in late log phase
Per team

Escherichia coli phage T4 suspension

9-ml tryptone blanks, 4

4-ml overlay agar tubes, 4

Tryptone agar base plates, 4

Sterile 1-ml pipets, 5

Procedure

First Session

1. Label four 9-ml tryptone blanks: 10^{-1}, 10^{-2}, 10^{-3}, 10^{-4} (figure 22.4).
2. Transfer 1 ml of the bacteriophage to the tube labeled 10^{-1} with a sterile 1-ml pipet. Discard the pipet. You must use fresh pipets each time so that you do not carry over any of the more concentrated phage to the next dilution.
3. Mix and transfer 1 ml of the 10^{-1} dilution to the 10^{-2} tube and discard pipet.
4. Mix and transfer 1 ml to the 10^{-3} tube and discard pipet.
5. Mix and transfer 1 ml to the 10^{-4} tube and discard the pipet.
6. Label four tryptone hard agar Petri plates: 10^{-1}, 10^{-2}, 10^{-3}, 10^{-4}.
7. Melt four tubes of soft overlay agar and place in a 50° C water bath. Let cool for about 10 minutes.
8. Add about 0.1 ml (or several drops) of *E. coli* broth to each tube of melted overlay agar.
9. Starting with the most diluted phage tube (10^{-4}), add 1 ml to the overlay agar and immediately pour on the tryptone agar base plate labeled 10^{-4}.
10. Using the same pipet, add 1 ml of the 10^{-3} dilution to a tube of overlay agar and pour into the plate labeled 10^{-3}. You can use the same pipet because you started with the most dilute sample and therefore the phage carried over are minimal.
11. Repeat for the 10^{-2} and 10^{-1} phage dilution.
12. Incubate the plates inverted at 37° C after the agar has hardened.

Second Session

1. Examine the plates. Select a plate containing between 30 and 300 plaques. As you count the plaques, place a dot with a marking pen under each colony on the bottom of the Petri plates. These marks can be wiped off so that each team member can count the colonies.
2. Estimate the numbers on the other plates. They should vary by a factor of 10 as the dilution increases or decreases.
3. To determine the titer, use this formula:

$$\text{No. of plaques} \times 1/\text{dilution} \times 1/\text{mls of sample} = \text{plaque forming units/ml.}$$

Example: If 76 plaques were counted on the 10^{-4} dilution, then:

$$76 \times 1/10^{-4} \times 1/1 = 76 \times 10^4 \text{ pfu/ml.}$$

4. Record results.

Figure 22.4 Schematic showing the procedure used to demonstrate *E. coli* phage plaques on the surface of agar plates. Plaques are represented by the light-colored areas on the 10^{-3} and 10^{-4} dilutions. (These results are only one possibility.)

Phage suspension

1.0 ml

1.0 ml   1.0 ml   1.0 ml

9 mls   9 mls   9 mls

10^{-1} 10^{-2} 10^{-3} 10^{-4} *E. coli* culture

Part A
Preparation of phage dilutions

1.0 ml   1.0 ml   1.0 ml   1.0 ml

0.1 ml   0.1 ml   0.1 ml   0.1 ml   0.1 ml

Part B
1. *E. coli* additions
2. Phage additions

Agar

10^{-1} 10^{-2} 10^{-3} 10^{-4} Control

Part C
Pouring of inoculated soft agar over plates of tryptone agar and incubating at 37°C inverted

10^{-1} 10^{-2} 10^{-3} 10^{-4} Control

Part D
Observation of plates for viral plaques

EXERCISE

22

Laboratory Report: Prokaryotic Viruses:
Bacteriophage Isolation and Titering

Results

1. Isolation and Culture from Raw Sewage

   a. Kinds of plaques _____ Number _____

   b. Did any plaques increase in size after reincubating? If yes, were small, very motile bacteria observed?

6.8×10^{-10} 1.18×10^{-10}

2. Titering a Phage Suspension

   a. How many plaques were on a countable plate? $10^{-9} \times 68$ $10^{-8} \times 118$

   b. Did you have about 10 times more plaques on the plate 10 times less dilute?

   c. How many phage per ml were in the original suspension? Show your calculations.

Questions

1. Why was the sewage sample filtered?

2. How can you distinguish a lytic phage from a temperate phage when observing plaques from the filtered sewage sample?

3. Why can a plaque be considered similar to a bacterial colony?

<u>Normal Flora</u> Upper Respiratory Tract

1) Take up space
2) Compete for nutrients
3) Synthesize important products (Vit X)
4) Produce antibacterial products
5) Keep immune system activated

Staph
Strep (α - hemolytic)

Diphtheroids
Spirochetes

β - Strep. pyogenes

Streptococcus

hemolysis
Lancefield
s.p. B-group B

H_2O_2 - catalase breaks down

$2H_2O_2 \xrightarrow{\text{catalase}} 2H_2O + O_2$

<u>catalase test</u>: differentiates between aerobic + nonaerobic organisms (differentiates between staph-strep)
                                                            +   -

bubbles indicate +

α hemolysis - partially break down red blood cells (green)
γ   "   — no
β   "   — complete (clear)

PART

THREE

Microbes and Humans

Enterobacteria
gram - rods

tell sheet book (unknown)

MR   VP   A α-naphthol 15 drops
methyl red   B (KOH) 15 drops

1) <u>Carbohydrate fermentation test</u> : glucose, sucrose, lactose + Durham tube
phenol red - pH indicator - red neutral → yellow, acid
acid is usually produced during sugar utilization
note whether or not gas was produced

both test   2) { <u>Methyl Red Test</u> : mixed acid   red : acid / clear : norm
for end
products   3) { <u>Voges Proskauer Test</u> : acetain   use 1 test tube

4) Indole test - red ring around top (+) - add indole
tryptophan

5) Citrate Test - utilize citrate as C source
enzyme - citrase   bromthymol blue = indicator
green agar in slant = acid change causes change to blue (+)

6) Urease - Breaks down urea - ammonia produced (alkaline)
pinkish color is positive

7) Motility test - use gelatin deeps - need inoculating needle - stab down
agar
growth out to sides (+)

40

*I**NTRODUCTION* to Medical Microbiology

It is easy to think of microorganisms as a deadly, vicious force—especially when the diseases they cause kill young people or wipe out whole populations. The organisms, however, are simply growing in an environment they find favorable.

If pathogens become too efficient at taking advantage of their host, the host dies and the organism dies with it. Thus, the most successful pathogens are those that live in balance with their host. When a new pathogen enters the population, it is very virulent, but after awhile there is a selection toward less virulent pathogens and also a selection in the hosts for increased resistance.

Medical microbiology continues to offer challenges to those interested in medicine and in pathogenic bacteria. These next exercises are an introduction to many of these organisms that are encountered in a clinical laboratory. Not only will you study the characteristics of the organisms, but also you will learn some strategies for isolating and identifying them. In addition, these exercises are designed to help you learn to differentiate between organisms you can expect to find as normal flora in various places in the body and others that are responsible for certain diseases.

After the aforesaid pestilence, many buildings both large and small, in all cities, towns and villages had collapsed and had completely fallen to the ground in the absence of inhabitants. Likewise, many small villages and hamlets were completely deserted: there was not one house left in them, but all those who had lived in them were dead. It is likely that many such hamlets will never again be inhabited. For all the beasts and cattle that a man possessed wandered about without a shepherd and everything a man had was with out a caretaker.

The chronical of Henry Knighton (1350): the impact of the Black Death

---

bromcresol purple - glucose - yellow (acid)

Mannitol Salt S. aureus & epidermitis - turns plate yellow (from pink)

PEA - selective

Ex. 4, 10 - 14, 15, 17, 18, 11, 12, 13, 20, 24, 25

EXERCISE 23

Normal Skin Flora

Getting Started

The organisms growing on the surfaces and in the orifices of the body are called normal flora. They are usually considered commensals because they do not harm their host, and, in fact, have several beneficial roles. Normal flora prevent harmful organisms from colonizing the skin because they are already established there and utilize the available nutrients. Some produce enzymes or other substances that inhibit nonresident organisms. Other organisms, called transients, can also be found on the skin for short periods, but they cannot grow there and soon disappear.

Familiarity with organisms making up the skin flora is useful because these organisms are frequently seen as contaminants. Skin is continually flaking off, and bacteria floating in the air on rafts of skin cells sometimes settle into open Petri dishes. If you are familiar with the appearance of *Staphylococcus* and *Micrococcus* colonies, for instance, you will be able to suspect contamination if you see such colonies on an agar plate. *Staphylococcus epidermidis* can also be seen in clinical specimens such as urine samples. These organisms probably are not causing disease, but are simply contaminants from skin flora.

Some of the organisms you may isolate:

Staphylococcus epidermidis   This gram-positive coccus is found on the skin as part of the normal flora of almost all humans throughout the world. It can also be isolated from many animals.

Staphylococcus aureus   At least 20% of the population "carry" (have as part of their normal flora) this bacterium. It is found on the skin, especially in the nares or nostrils, and it seems to cause no harm to its host. However, *S. aureus* is frequently the cause of wound infections and food poisoning, and has been implicated as the cause of toxic shock syndrome. It can be identified by the coagulase test. Recently, many additional species of

Staphylococcus have been identified. They are associated mostly with diseases in immunologically compromised individuals.

Micrococcus luteus   This is a gram-positive coccus found on the skin of some people, but it almost never causes disease. It is frequently an air contaminant forming bright yellow colonies.

Propionibacterium acnes   These anaerobic, gram-positive rods are diphtheroid or club shaped. When investigators tried to isolate an organism that might be the cause of acne, they almost always found the same gram-positive diphtheroid rods in the lesions. Therefore, they named the organism *Propionibacterium acnes*. However, when people without acne were studied, it was found that *P. acnes* was present on their foreheads as well. Although some people have a much higher population of these organisms than others, the number of organisms does not seem to correlate with acne or any other skin condition.

Propionibacterium granulosum   A gram-positive diphtheroid rod found on some individuals, usually in smaller numbers than *P. acnes* (figure 23.1). It is considered a harmless commensal.

Definitions

Coagulase test.   A test in which organisms are mixed with plasma on a slide. If the cells clump together, the culture is coagulase positive.

Commensals.   Organisms that live together in close association and may or may not benefit each other.

Diphtheroid.   A gram-positive, club-shaped organism sometimes called a coryneform. *Propionibacterium* and *Corynebacterium* are examples of diphtheroid-shaped organisms.

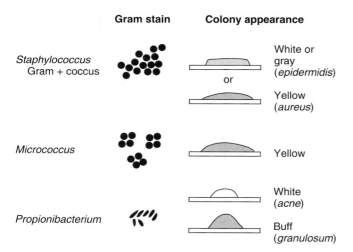

	Gram stain	Colony appearance	
Staphylococcus Gram + coccus		White or gray (*epidermidis*) or Yellow (*aureus*)	
Micrococcus		Yellow	
Propionibacterium		White (*acne*) Buff (*granulosum*)	

Objectives

1. To learn to identify organisms making up the normal skin flora.
2. To understand the importance of skin flora.
3. To learn about the anaerobe jar.

References

Marples, Mary J. "Life on the human skin." *Scientific American*, January 1969, 220(1):108–115.
Nester et al. *Microbiology: A human perspective.* Chapter 19, 421–424.
Safety Precautions: Some students may isolate *Staphylococcus aureus* as part of their normal flora. This is a pathogen and should be handled with extra care.

Procedure

First Session

1. Saturate a cotton ball with 95% alcohol and rub the forehead for 20 seconds. This will remove any transient organisms you might have on your skin, which are not part of your normal flora.
2. Let the forehead dry for about 20 minutes. Avoid touching it with your hair or fingers.

Materials

First Session

TSY agar plates (or TSY contact plates 2 inches in diameter), 2

GasPak anaerobe jar (or other anaerobic system)

Sterile swab

Sterile saline

95% ethanol

Cotton balls

Second Session

Tubes of sterile water (0.5 ml/tube), 6

TSY + glucose agar deeps (yeast extract and glucose are added to TS agar to encourage the growth of *Propionibacterium*), 4

Glycerol + brom cresol purple agar slants, 2

Magnifying glass is optional but helpful

Third Session

Plasma

3. Moisten a sterile swab with saline and rub it briskly on an area of your forehead about the size of a quarter for about 15 seconds.
4. Immediately swab the first third of a TSY agar plate, discard the swab, and finish the streak plate with a loop.
5. Repeat the procedure, swabbing a second TSY agar plate from an adjacent area of the forehead.

   (Alternative method to steps 3, 4, and 5: Press an open contact TSY agar plate on the forehead. Repeat with a second plate on an adjacent area.)
6. Incubate one of the TSY agar plates aerobically at 37° C. Incubate the second TSY agar plate anaerobically in a GasPak or other anaerobe jar at 37° C. Follow the manufacturer's directions for creating an anaerobic atmosphere.

7. After 48 hours of incubation, the student or instructor should store the aerobic plate at room temperature to prevent the plate from drying out. *Staphylococcus* and *Micrococcus* can be observed after 48 hours, but *Propionibacterium* must be incubated five days before colonies can be seen.

Second Session

(5 days later)

1. Examine the aerobic TSY plate and circle two different colony types with a marking pen on the bottom of the plate. Make a Gram stain of each circled colony.
2. If the colonies are gram-positive cocci, suspend the remainder of the colony used for the Gram stain in 0.5 ml sterile saline. Use this suspension to inoculate:
   a. a glucose + brom cresol purple TS agar slant
   b. a cooled melted agar deep (sometimes called a shake tube). See exercise 9 for inoculation procedure.
3. Examine the plate incubated in the anaerobe jar. You will see some of the same colony types observed on the aerobic plate, because *Staphylococcus* are facultative anaerobes and can grow with or without oxygen. *Propionibacterium* colonies, however, are white and very small—only a few millimeters in diameter. *P. granulosum* are slightly larger and appear as buff or pink cones when the plate is examined with a magnifying glass. Choose two possible *Propionibacterium* colonies, Gram stain them, and if they are diphtheroid gram-positive rods, inoculate into a shake tube. You can inoculate the shake tube directly from the plate if very little remains of the colony.

Third Session

1. Observe the glucose + brom cresol slants. If the organism is able to ferment glucose, the tube will turn from purple to yellow.
2. Observe the agar deeps. Obligate aerobes are only able to grow on the top, while facultative anaerobes will grow throughout the entire tube. The obligate anaerobes will not be able to grow in the top few centimeters where oxygen has diffused in—only in the bottom anaerobic portion. (See exercise 9.)
3. Identify your isolates. The following is a description of the organisms most commonly found on the forehead.

Staphylococcus are gram-positive cocci arranged in clusters. They are facultative and can ferment glucose. There are two main species of *Staphylococcus* found on the skin: *epidermidis* and *aureus*. *S. aureus* tends to have yellow colonies and *S. epidermidis* white colonies.

Optional: If you have yellow colonies of *Staphylococcus*, you can determine if it is *S. aureus* with a coagulase test. *S. aureus* is coagulase positive and *S. epidermidis* is coagulase negative.

Coagulase Test

Place a drop of water on a slide and make a *very* thick suspension of cells from a yellow colony.

Place a drop of plasma next to it and mix the two drops together. Look for clumping; clumped cells indicate a coagulase-positive result.

Drop the slide in boiling water and boil for a few minutes to kill the organisms before cleaning the slide.

Micrococcus are gram-positive cocci arranged in packets of four or eight. They are obligate aerobes and cannot ferment glucose or other sugars.

Propionibacterium acnes are diphtheroid coryne-form gram-positive rods that tend to palisade or line up like a picket fence. They form very small, white, glistening colonies. *Propionibacterium* are obligate anaerobes that grow only in the absence of oxygen, but are not killed by air as are some anaerobes.

Propionibacterium granulosum appears the same as *P. acnes* in a Gram stain but forms pinkish, slightly larger colonies.

4. Record your results.

Summary of Reactions			
	Gram stain	**Metabolism**	Glucose
Staphylococcus epidermidis	+ cocci	facultative	+
Staphylococcus aureus	+ cocci	facultative	+
Micrococcus	+ cocci	obligate aerobe	–
Propionibacterium	+ rods	obligate anaerobe	

coagulase   Impetigo

fibrinogen → fibrin
clot-like
meshwork

EXERCISE

23

Laboratory Report: Normal Skin Flora

Results

	Isolate 1	Isolate 2	Isolate 3	Isolate 4
Gram stain				
Colony appearance				
Glucose fermentation				
Agar deep				
Possible identity				

Questions

1. How could normal skin flora be helpful to the host?

2. Why did you wipe your forehead with alcohol before sampling it?

3. Why was *Staphylococcus* the only organism that could grow on both plates?

4. Why does *Staphylococcus* probably cause more contamination than *Propionibacterium*, even though most people have higher numbers of the latter?

EXERCISE

24

Respiratory Microorganisms

Getting Started

In this exercise, you have an opportunity to observe a throat culture and learn some of the ways a clinical microbiologist identifies pathogenic organisms. You also examine the normal flora of the throat, and while you will not work with actual pathogens, many of the organisms in the normal flora resemble related pathogenic bacteria. In exercises 30 and 31, you use serological techniques to further differentiate some of these organisms.

A physician frequently orders a throat culture if a patient has a very sore throat and fever. This is done to determine whether or not the sore throat is caused by Group A β-hemolytic *Streptococcus pyogenes*. This organism is important because it not only causes strep throat, a disease characterized by fever and a very sore throat, but also two very serious diseases can follow the original infection: rheumatic fever, a disease of the heart, and acute glomerulonephritis, a disease of the kidney.

Fortunately, streptococci are usually still sensitive to penicillin and related antibiotics, and treatment is fairly straightforward. However, most sore throats are caused by viruses. Since viruses do not have a cell wall or any metabolic machinery, they are not affected by penicillin or most other antibiotics. Therefore, it is important to make an accurate diagnosis so that antibiotics can be used wisely.

Other pathogens sometimes found in the throat are *Streptococcus pneumoniae*, *Neisseria meningitidis*, *Haemophilus influenzae*, and *Klebsiella pneumoniae*. If these organisms are indeed causing disease, they are usually present in large numbers and the patient has the symptoms of the disease.

β-hemolytic *Streptococcus* can be identified when growing on blood agar. This medium is made of a base agar that is rich in vitamins and nutrients. Before pouring the melted agar into the Petri plates, 5% sheep blood is added. (Sheep raised for this purpose donate blood as needed.) The blood performs two functions: it adds additional

nutrients and aids in distinguishing α-hemolytic from β-hemolytic streptococci.

Streptococci produce hemolysins that act on red blood cells. α-hemolytic streptococci, which are a major component of the normal throat flora, incompletely lyse red blood cells. When the organism is growing on blood agar, a zone of partial clearing of the red blood cells can be seen around a colony. β-hemolytic streptococci produce hemolysins that completely lyse the red blood cells and therefore produce a clear zone in the blood agar around the colony. It is important to note that β-hemolysis is not always correlated with pathogenicity. For example, some strains of *E. coli* can produce β-hemolysis but are not responsible for any disease.

Commercial kits based on specific antibodies or other methods are now available that can be used to determine if a patient has strep throat. Although these tests can be performed in the doctor's office, the results sometimes must be verified by growing the culture on blood agar.

A small percentage of the hemolysins of β-hemolytic streptococci are oxygen labile, which means they are destroyed by oxygen. In a clinical laboratory, throat cultures are incubated in an anaerobe jar so that no hemolysis is overlooked. In this laboratory exercise, we will incubate the cultures in a candle jar, if convenient. This increases the CO_2 content of the atmosphere and enhances the growth of some organisms, but does not create anaerobic conditions.

The throat contains a plethora of organisms that make up the normal flora, many of which resemble pathogenic organisms. Frequently, some actual pathogenic organisms are found in small numbers, such as *Streptococcus pneumoniae* or β-hemolytic *Streptococcus*. The presence of these organisms is only significant when they appear in large numbers.

Examples of common normal flora of the throat are diagrammed in figure 24.1. You might observe some of them on your streak plate.

Figure 24.1 Normal flora of the throat.

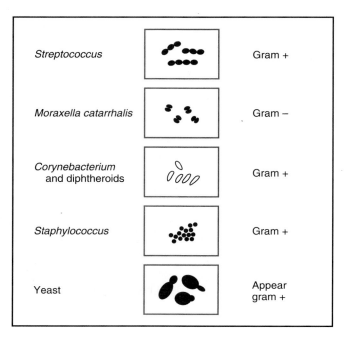

Streptococcus		Gram +
Moraxella catarrhalis		Gram −
Corynebacterium and diphtheroids		Gram +
Staphylococcus		Gram +
Yeast		Appear gram +

α-hemolytic streptococcus These organisms will be the predominant organisms growing on your plates. Another name for them is Viridans streptococci, and they rarely cause disease. They are gram positive and grow in short chains. They are catalase-negative, which distinguishes them from *Staphylococcus*. *Streptococcus pneumoniae* is also α-hemolytic and is differentiated from the normal flora by sensitivity to certain antibiotics and serological tests. They are all oxidase negative.

Moraxella catarrhalis These organisms are gram-negative cocci arranged in pairs. They resemble the microscopic morphology of *Neisseria* (which grows only on a different kind of heated blood agar called chocolate agar). The genus *Neisseria* includes such pathogens as *N. gonorrhoeae* and *N. meningitis*. Try to find a colony of *Moraxella* and Gram stain it so you can be familiar with the appearance of these organisms. The colonies may be a little larger than *Streptococcus* and are oxidase positive, so if oxidase reagent is available, it will help you identify a colony. They are also catalase positive.

Note: *Moraxella catarrhalis* was formerly named *Branhamella catarrhalis*, and prior to that *Neisseria catarrhalis*.

Corynebacterium and Diphtheroids These are irregular, club-shaped gram-positive rods. They are part of the normal flora but resemble *Corynebacterium diphtheriae*, which causes diphtheria. They are usually catalase positive and oxidase negative.

Staphylococcus These gram-positive cocci are arranged in clusters. *Staphylococcus aureus* frequently is part of the normal flora, although it is a potential pathogen. The colonies are usually yellow. *Staphylococcus* is catalase positive and oxidase negative.

Yeasts These are fairly common in the oral flora and they form relatively large colonies. In a Gram stain, the cells (which are eukaryotic) appear purple, are larger than bacteria, and sometimes have buds. They are catalase positive and usually oxidase positive.

Definitions

Catalase. An enzyme found in most aerobic organisms that breaks down H_2O_2 to water and oxygen.

Oxidase. A reagent that tests for cytochrome C.

Serological test. Identification of organisms by mixing cells with serum containing antibodies to a specific organism. If the cells clump, the test is positive for that organism.

Objectives

1. To learn the importance of Group A β-hemolytic *Streptococcus* and how to distinguish it from normal flora.
2. To observe normal flora of the throat.

References

Nester et al. *Microbiology: A human perspective.* Chapter 20, 450–460.

Ryan, Kenneth J., ed. *Sherris medical microbiology, an introduction to infectious diseases,* 3d ed. Norwalk: Appelton and Lange, 1994.

Sneath, Peter H. A., ed. *Bergey's manual of systematic bacteriology.* Baltimore: Williams & Wilkins, 1986.

Safety Precautions: There may be colonies of β-hemolytic streptococci (*Streptococcus pyogenes*) and *Staphylococcus aureus* on the agar plates of normal flora. Handle these plates and the demonstration plates of β-hemolytic *streptococcus* with care.

Materials

Per student
First Session

  Blood agar plate, 1

  Sterile swab, 1

  Tube of sterile saline, 1

  Tongue depressor

  Demonstration cultures

    α-hemolytic *streptococcus* on blood agar

    β-hemolytic *streptococcus* on blood agar

  Candle jar

Second Session

  H_2O_2

  Oxidase reagent

Procedure

First Session

1. Swab your partner's throat. First seat your partner on a stool and shine a light on the throat. Carefully remove a sterile swab from the wrapper and moisten with the sterile saline, pressing out excess moisture on the inside of the tube. Depress the tongue with the tongue depressor and swab the tonsilar area on the side of the throat (figure 24.2). Do not swab the hard palate directly in the back behind the uvula and do not touch the tongue or lips. Do this rather quickly to avoid the gag response.
2. Swab the first third of a streak plate on the blood agar plate, rolling the swab over the agar to be sure to inoculate all sides. Discard the swab and continue streaking the rest of the plate with your loop as usual for isolated colonies.
3. Incubate your plate at 37° C in a candle jar, if available.

Figure 24.2 Diagram of open mouth. Shaded areas indicate places to swab. From Josephine A. Morello, Helen Eckel Mizer, and Marion E. Wilson, *Laboratory Manual and Workbook in Microbiology: Applications to Patient Care.* Copyright © 1994 Wm. C. Brown Communications, Inc., Dubuque, Iowa. All Rights Reserved. Reprinted by permission.

Second Session

1. Examine your throat culture plate and compare it to the pure cultures of α- and β-hemolytic streptococci on the demonstration plates. If you observe β-hemolysis, make a Gram stain of the colony to determine if it is a gram-positive streptococcus. If they indeed appear to be β-hemolytic streptococcus colonies, are they the predominant colony type on the plate?
   In a clinical laboratory:
   a. If they were the most numerous colony type, the physician would be notified and the patient would be treated.
   b. If only a few colonies were present, then the results could be reported as "ruled out β strep."
2. Make Gram stains of various colonies and do catalase and oxidase tests on the same colonies if possible. Record the kinds of organisms you observed based on their Gram stain, colony morphology, catalase test, and oxidase test. See Getting Started for a description of some of the organisms you might see.

Figure 24.3 (*a*) Catalase and (*b*) oxidase tests.

(a)   Bubbles in H_2O_2 indicate a positive catalase test.

Labels: Cells, H_2O_2, Glass slide

(b)   Cells rubbed on paper turn dark pink indicating positive oxidase test.

Labels: Paper saturated with oxidase reagent, Glass slide

Tests

Catalase. With a sterile loop, place some cells from the colony to be tested on a glass slide. Cover the cells with a few drops of H_2O_2. If bubbles are formed, the culture is catalase positive (figure 24.3). Boil the slide in water for a few minutes to kill the organisms.

Oxidase. Place a small piece of filter paper on a glass slide and moisten with freshly prepared oxidase reagent. Remove some cells from a colony to be tested with a sterile loop and rub the loop on the moistened filter paper. If a pinkish-purple color appears, the cells are oxidase positive (see figure 24.3). Place the paper in an autoclave bag and boil the slide for a few minutes to kill the organisms.

EXERCISE

24

Laboratory Report: Respiratory Microorganisms

Results

	Appearance of Hemolysis	Numbers Present
α-hemolytic streptococci		
β-hemolytic streptococci		

Indicate the numbers of organisms, using ++++ for the most numerous or highest number, and + if very few are observed. Use ++ and +++ for numbers in between.

Other organisms observed:

Possible Identity	Gram Stain	Catalase	Oxidase

Questions

1. What is the predominant organism in your throat flora?

2. What is the difference between alpha- and beta-hemolysis?

3. Did you observe any β-hemolytic streptococci in your throat?

4. How many different organisms did you observe in your throat culture?

EXERCISE 25

Identification of Enteric Gram-negative Rods

Getting Started

In this exercise, you will learn how to identify bacteria with biochemical tests. Since it is frequently necessary to identify gram-negative rods in the clinical laboratory, these are excellent organisms to use as examples of the general procedure. Enteric organisms, for instance, are part of the normal flora of the intestine, but are responsible for the majority of the urinary tract infections. At times, organisms from the colon (large intestine) can infect the bladder and grow in the urine stored there.

Biochemical tests measure such things as the differences in the organisms' ability to ferment different sugars, the presence of various enzymes, and physical characteristics such as motility. An organism then can be identified by comparing the results of the tests to the results of known bacteria in such reference books as *Bergey's Manual of Systematic Bacteriology*. The following is a discussion of the tests you will be using and how they work.

Fermentation Tubes Carbohydrate fermentation tubes consist of a complete broth, a carbohydrate such as the sugar glucose, and the pH indicator andrades. A small Durham tube is added, which fits inside the large tube. If the organism can ferment the sugar, it will produce acidic products and the pH will fall, turning the pH indicator pink. If the organism also produces a gas such as hydrogen, some of it will accumulate in the Durham tube as a bubble (see color plate 21).

After incubation, you should examine the tubes for growth (sometimes you have to shake the tubes slightly because the cells have settled). If for some reason there is no growth, the test must be repeated. If there is growth, the results can be recorded as:

A   Acid production, if the indicator has turned red

A/G   Acid and Gas, if the indicator has turned red and a bubble is seen in the Durham tube

N/C   No Change, if neither gas nor acid has formed

The sugars glucose, sucrose, and lactose are especially useful in the identification of the enteric gram-negative rods. Lactose fermentation aids in the preliminary differentiation of enteric pathogens from the normal coliforms. *Salmonella* and *Shigella* species, which cause enteric diseases of various kinds, do not ferment lactose; however, members of the common fecal flora, *Escherichia coli*, *Enterobacter aerogenes*, and *Klebsiella* are able to ferment lactose. Lactose fermentation is not related to pathogenicity in any way, it is simply a convenient characteristic for identifying organisms. The enteric *Proteus*, for example, is lactose negative, but usually a nonpathogen.

Methyl Red and Voges-Proskauer The fermentation tubes previously described show whether or not fermentation has taken place. On the other hand, these two tests identify a particular kind of fermentation.

Fermentation is the energy-yielding pathway a facultative organism can use if oxygen is unavailable. Organisms have different fermentation pathways resulting in different end products. Some are simple, such as the conversion of pyruvate to lactic acid. Other pathways are more complex, yielding a variety of products and perhaps additional energy (figure 25.1).

Methyl Red This test measures the pH after organisms have grown in buffered peptone glucose broth (MR/VP broth). It is positive when the pH is less than 4.3. *E. coli* and other organisms ferment sugars by the mixed acid pathway. The products of this pathway are predominantly acetic and lactic acids, some organic compounds, and CO_2 and H_2. This results in a very low pH and therefore *E. coli* is methyl red positive.

Figure 25.1 Fermentation pathways.

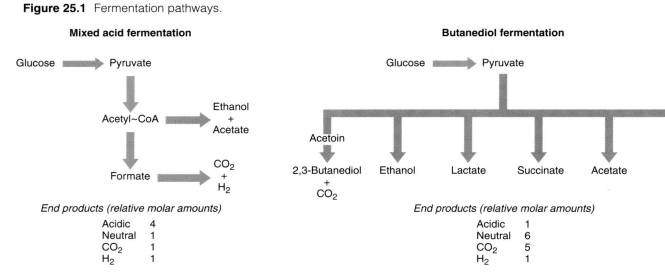

Mixed acid fermentation

Glucose ⟹ Pyruvate

Acetyl~CoA ⟶ Ethanol + Acetate

Formate ⟶ CO_2 + H_2

End products (relative molar amounts)

Acidic   4
Neutral   1
CO_2 1
H_2 1

Butanediol fermentation

Glucose ⟹ Pyruvate

Acetoin

2,3-Butanediol + CO_2 Ethanol   Lactate   Succinate   Acetate CO_2 + H_2

End products (relative molar amounts)

Acidic   1
Neutral   6
CO_2 5
H_2 1

Other bacteria such as *Enterobacter* use the butanediol pathway when fermenting sugars. The end products are predominantly alcohols and a small amount of acid, as well as CO_2 and H_2. Although the pH is low, it is not low enough to change the methyl red, so these bacteria are methyl red negative.

Voges-Proskauer   One of the intermediates in the butanediol pathway is acetoin, for which the Voges-Proskauer reagents test. If the test is positive, then acetoin is present and the organism is using the butanediol pathway. The reagents Voges-Proskauer A and B are added to the broth culture after incubation. If a brick red precipitate forms, it is VP +.

Indole   Some organisms have an enzyme that cleaves the amino acid tryptophan, producing indole. They can be grown in tryptone broth, which has a high level of tryptophan. After incubation, the broth is tested for the presence of indole by adding Kovacs reagent. A red ring forms on the top of the broth if the organism is indole +.

Citrate Utilization   The organism is grown on Simmons citrate medium, a mineral medium with citrate as the sole carbon source. The agar also contains the pH indicator bromthymol blue. If the organism can utilize the citrate, the pH rises and the indicator turns a deep blue.

Urea Hydrolysis   Organisms are grown on agar containing urea and a pH indicator. If the organism produces the enzyme urease, urea is split, forming ammonia and CO_2. This raises the pH of the medium, turning it bright pink. *Proteus* species can be distinguished from other enteric rods because it always produces urease.

Motility   Organisms are stabbed into a semisolid agar deep containing tetrazolium, an electron acceptor that turns red wherever there is growth. After incubation, a diffuse red color throughout the medium indicates motility. If there is a red streak only where the agar was stabbed, then the organism is nonmotile.

In this exercise, each team inoculates a series of biochemical tests with a labeled organism and an unlabeled "unknown" organism. (The "unknown" is one of the organisms listed.) There may seem to be a large number of tubes to inoculate, but if they are labeled and lined up in a test tube rack, inoculation can be done fairly quickly. Good organization is very helpful.

Note: Commercial test systems (such as *Enterotube II System* from Becton Dickinson) are available in which a whole series of tests are inoculated at the same time (see color plate 22). After incubation, the results are read and the organism can be identified.

Definitions

Enteric. Associated with the intestine.

Fermentation. An energy-yielding metabolic pathway in which organic compounds are both electron donors and acceptors.

Coliforms. Gram-negative rods found in the intestine that ferment lactose—*Escherichia coli*, *Enterobacter*, and *Klebsiella*.

Objectives

1. To learn to identify bacteria using biochemical tests.
2. To understand the physiological basis for the tests.
3. To become familiar with organisms commonly seen in a clinical laboratory, especially those causing urinary tract infections.

References

Nester et al. *Microbiology: A human perspective*. Chapter 24, 544–547.

Materials

Per team of two students
Cultures on TS agar slants of

  Escherichia coli

  Enterobacter aerogenes

  Klebsiella pneumoniae

  Pseudomonas aeruginosa

  Proteus mirabilis

Sterile saline 1.0 ml/tube, 2 tubes

Glucose fermentation tubes, 2

Lactose fermentation tubes, 2

Sucrose fermentation tubes, 2

Tryptone broths, 2

Methyl Red-Voges-Proskauer (MR-VP) broth tubes (one tube for each test), 4

Simmons citrate slants, 2

Motility agar deeps, 2

Reagents

  Oxidase reagent (freshly prepared) and small squares of white filter paper

  Kovacs Reagent

  Methyl red

  Voges-Proskauer reagents A and B

Procedure

First Session

1. Choose one labeled culture and one "unknown" culture (or the cultures you are assigned).

2. Label a set of tubes for each organism with the name of the organism, the medium, the date, and your name or initials. Note that you will need 1 MR-VP broth for the methyl red test and another MR-VP broth for the Voges-Proskauer test. You should have a total of eight tubes for each organism to be identified.

3. Make a suspension of each organism by adding bacteria with a sterile loop to the sterile saline. Use this suspension to inoculate the test media.

4. Inoculate the tubes by adding a loopful of the suspended organisms to the broth. Inoculate the agar slant by gliding the loop over the surface of the agar, starting at the bottom of the slant. Use good aseptic technique to avoid contamination with unwanted organisms.

5. Inoculate the motility deep by stabbing the agar almost to the bottom with your inoculating loop. Use cells from the original slant instead of from the suspension to be sure you have enough cells.

6. Perform an oxidase test using cells from the slant. See exercise 24.

7. Incubate the tubes at 37° C for 2 days or more. Incubate the motility deeps at room temperature. Some organisms are not motile at 37° C.

Second Session

1. Examine the fermentation tubes and record as A if the pH indicator has turned red from acid production, G for gas production, and N/C for no change if neither acid nor gas has formed. Be sure there is growth in the tube before recording.

2. Add a dropper full of Kovacs reagent to the tryptone broth and shake slightly. A red layer on the top of the broth indicates a positive test for indole.

3. Add a few drops of methyl red to one of the MR-VP broths. A red color indicates a positive test.

4. Add 10–15 drops of V-P reagent A (alpha-naphthol solution) and 10–15 drops of V-P reagent B (40% KOH). Shake and let stand a few minutes, or an hour maximum. The appearance of a red color indicates a positive test.

Safety Precaution: Alpha-naphthol is toxic.

5. Examine the motility deep. If the tube appears pink throughout the agar, it is positive for motility. If only the original stab line appears pink, the test is negative for motility.

6. Record the results of your known organism in the chart below and on a similar chart drawn on the blackboard. These will be the results to establish the reactions of the known bacteria. Consult your instructor if student results of their known bacteria do not agree.

7. Record the class results on the board in the results section. Also record the results of your unknown organism.

	Fermentation								
	glu	lac	suc	indole	MR	VP	cit	mot	ox
E. coli									
Enterobacter									
Klebsiella									
P. mirabilis									
Pseudomonas									

EXERCISE

25

Laboratory Report: Identification of Enteric Gram-negative Rods

Results

	Fermentation			indole	MR	VP	cit	mot	ox
	glu	lac	suc						
E. coli									
Enterobacter									
Klebsiella									
P. mirabilis									
Pseudomonas									
Unknown									

Questions

1. What is the identity of your unknown organism?

2. Can you determine whether an organism can ferment a sugar if it does not grow in the broth? Explain.

3. How can an organism have a positive test for acid from glucose in a fermentation tube, but have a negative methyl red test, which is also a glucose fermentation test?

4. Were there any organisms that did not ferment any sugars? If yes, which organisms?

EXERCISE

Clinical Unknown Identification

Getting Started

In this exercise, you have an opportunity to utilize the knowledge and techniques you have learned in order to identify a mixture of two unknown organisms. You are given a simulated (imitation) clinical specimen containing two organisms, and your goal is to separate them into two pure cultures and identify them using different media and tests. The organisms are either associated with disease or are common contaminants found in normal flora or the environment.

Your unknown specimen represents either a urine infection or a wound infection. In actual clinical cases, standardized procedures exist for each kind of specimen. However, you will be identifying only a limited number of organisms. With some careful thought, you can plan logical steps to use in identifying your organisms.

The following are characteristics useful in identifying your unknown organism.

Bacterial Cell Morphology   The size, shape, arrangement, and Gram staining characteristics of the bacteria as determined by the Gram stain. It also could include the presence of special structures such as endospores.

Colonial Morphology   The appearance of isolated colonies on complete media such as TS agar or blood agar, including their size, shape, and consistency.

Growth on Selective Media   The ability of organisms to grow on selective media. Mannitol salt (selects for organisms tolerating 7.5% salt), EMB (selects for gram-negative organisms), and MacConkey (selects for gram-negative rods).

Reactions on Differential Media   The color of colonies on eosin methylene blue agar (EMB) or MacConkey agar is based on lactose fermentation. The appearance of organisms on mannitol

salt agar is based on mannitol fermentation. The presence of hemolysis on blood agar constitutes another type of reaction.

Biochemical Capabilities   The ability to ferment different carbohydrates and the production of various end products. The formation of indole from tryptophan and tests such as methyl red, Voges-Proskauer, citrate utilization, urease, catalase, oxidase, and coagulase.

Approach the identification of your "unknown" clinical specimen with the following steps:

1. Make a Gram stain of the specimen.
2. Streak the broth on a complete medium and a selective medium that seem appropriate.
3. After incubation, identify two different colony types and correlate with their Gram reaction and shape. Also correlate the growth and appearance of the colonies on selective media with each of the two organisms.
4. Restreak for isolation. It is useless to do any identification tests until you have pure cultures of the organisms.
5. After incubation, choose a well-isolated colony and inoculate a TS agar slant to be used as your stock culture. Prepare a stock culture for each organism.
6. Inoculate or perform various tests that seem appropriate. Keep careful records. Record your results on the worksheets as you observe them.
7. Identify your organisms from the test results.

Definitions

Clinical specimen.   Cultures encountered in a medical laboratory.

Differential media.   Media that permit the identification of organisms based on the appearance of their colonies.

Selective media. Media that permit only certain organisms to grow and aid in isolating one type of organism in a mixture of organisms.

Objectives

1. To give you an opportunity to apply your knowledge to a microbiological problem.
2. To give you insight into the procedures used to isolate and identify clinical specimens.
3. To teach you to be aware of the presence of contaminants or nonpathogens in clinical specimens.

Materials

First Session
Unknown mixture labeled with hypothetical source (for each student or team of two students)

  Blood agar plate or TS agar plate, 1 per student

  MacConkey agar plate (or EMB agar plate)

  Mannitol salt agar plate

Second and Third Sessions

  TS agar plates

  Nutrient agar slants

  Citrate agar slants

  Urea slant

  Glucose + bromcresol purple agar slants

  Fermentation broths of glucose, lactose, sucrose

  MR-VP broth for the Voges-Proskauer and methyl red test

  Tryptone broth for the indole test

  Kovacs reagent

  Voges-Proskauer reagents A and B

  Methyl red for methyl red test

  Plasma for coagulase test

  Staining material for endospores and capsules

  H_2O_2

References

Difco Manual of Dehydrated Culture Media. Difco Laboratories, 10th ed. Detroit, 1984.

Holt, John G., editor-in-chief. *Bergey's manual of systematic bacteriology.* Volumes 1 and 2. Baltimore: Williams & Wilkins, 1984.

Nester et al. *Microbiology: A human perspective.* Chapter 24, 545–546; chapter 26, 597–608.

Procedure

First Session

1. Make a Gram stain of the broth culture. Observe it carefully to see if you can see both organisms. You can save the slide and observe it again later if you have any doubts about it. You can also save the broth, but one organism may overgrow the other.
2. Inoculate a complete medium agar plate such as TS agar or blood agar, and appropriate selective and differential agar plates. Use MacConkey agar (if you suspect the possibility of a gram-negative rod in a urine specimen) or a mannitol salt agar plate (if you suspect *Staphylococcus* in a wound infection). Streak the plates for isolated colonies.
3. Incubate at 37° C.

Second Session

1. Examine the streak plates after incubation and identify the two different colony types of your unknown organisms either on the complete medium or the selective media, wherever you have well-isolated colonies. Gram stain each colony type (organisms usually stain better on nonselective media). Also identify each colony type on the selective and differential media so that you know which organisms can grow on the various media. Record their appearance on the differential media as well. It is helpful to circle colonies that you Gram stain on the bottom of the Petri plate with a marking pen.
2. Restreak each organism on a complete medium (instead of selective media) for isolation. This technique ensures that all organisms will grow and you will be able to see if you have a mixed

culture. Do not discard your original streak plates of your isolates. If at some point your isolate does not grow, you will be able to go back to the old plates and repeat the test.

Third Session

1. Observe the plates after incubation. If your organisms seem well isolated, inoculate each one on a TS agar slant to use as your stock culture. If you do not have well-isolated colonies, restreak them. It is essential that you have a pure culture. Possible steps in identifying gram-positive cocci follow.

2. Look at the possible list of organisms and decide which ones you might have based on the information you have found so far.

Gram-positive Cocci

Staphylococcus aureus Found in either urine or wounds	Gram-positive cocci in bunches, catalase positive, yellow colony, ferments glucose and mannitol, coagulase positive.
Staphylococcus epidermidis Contaminant	Gram-positive cocci in bunches, catalase positive, ferments glucose but not mannitol, coagulase negative.
Micrococcus Contaminant	Gram-positive cocci in packets, catalase positive, yellow colony, does not ferment glucose or mannitol, coagulase negative.
Enterococcus faecalis Found in urine	Gram-positive cocci in chains, catalase negative, glucose positive, coagulase negative.

Gram-positive Rod

Bacillus Contaminant in wounds	Large, gram-positive rods, forms spores, catalase positive.

Gram-negative Rods

Escherichia coli Urine	Glucose and lactose positive (acid and gas), indol positive, methyl red positive, Voges-Proskauer negative, citrate negative, urea negative, oxidase negative.
Proteus Urine	Lactose negative, oxidase negative, urea positive, glucose positive.
Pseudomonas Urine and wounds	Lactose negative, glucose negative, urea negative, oxidase positive.
Enterobacter	Glucose and lactose positive (acid and gas), indol negative, methyl red negative, Voges-Proskauer positive, citrate positive, urea negative, oxidase negative.

These are a few of the test results. Your instructor may provide more. Plan work carefully and do not waste media using tests that are not helpful. For example, a urea slant would not be useful for distinguishing between *Staphylococcus epidermidis* and *S. aureus*. Possible organisms included in unknowns:

Simulated wounds
 Staphylococcus epidermidis
 Staphylococcus aureus

 Micrococcus luteus
 Pseudomonas aeruginosa
Simulated urine infection
 Escherichia coli
 Enterobacter aerogenes
 Proteus
 Enterococcus faecalis
 (plus wound organisms)

This is just a partial list. Others can be added as desired.

NOTES:

Name _____ Date _____ Section _____

Partner _____ Unknown # _____ Source _____

EXERCISE

26

Results Final Worksheet:
Clinical Unknown Identification

Gram stain (original specimen) _____

_____

(describe cell shape, arrangement, and Gram stain)

Gram stain (TS agar subculture) _____

_____

(describe cell shape, arrangement, and Gram reaction)

Test	Organism #1	Organism #2
Colony description		
(Trypticase Soy agar or blood)		
Gram stain		
Colony appearance MacConkey (or EMB)		
Colony appearance mannitol salt agar		
Special stains		
capsule		
endospore		
Lactose fermentation		
Glucose fermentation		
Sucrose fermentation		
Mannitol fermentation		

Test	Organism #1	Organism #2
Indole production		
Methyl red		
Voges-Proskauer		
Citrate utilization		
Urea hydrolysis		
Motility		
Catalase test		
Coagulase		
Oxidase		

Final Species Identification:

Final Report

1. What is the identification of your organisms? Discuss the process of identification (reasons for choosing specific tests, any problems, and the like).

Organism #1:

Organism #2:

FOUR

Immunology

Introduction to Some Immunological Principles and Techniques

Immunology, the study of the body's immune response, is responsible for protecting the body against disease. It is often triggered when foreign substances or organisms invade the body. Examples include pathogenic microbes and chemical compounds they produce, such as foreign materials called **antigens.** In some diseases such as AIDS and cancer, the body's immune response is either seriously weakened or destroyed, whereas for milder diseases the immune response remains complete.

Different forms of the immune response include **phagocytic cells** such as white blood cells (WBCs), **enzymes** such as **lysozyme,** and **antibodies.** WBCs and enzymes are examples of **natural immunity** since they are already present in the body and need not be triggered by the antigen. In contrast, antibodies are an example of **acquired immunity** since their formation is triggered only in the presence of the antigen. For some people, the immune response can be triggered by the body's own proteins. This can result in the formation of **autoimmune diseases** such as rheumatoid arthritis and glomerulonephritis.

Phagocytic cells and enzymes are also an example of **nonspecific immunity** since they can react with a variety of different foreign substances (for instance, phagocytic cells can engulf both inanimate and animate particles). Conversely, antibodies represent a form of **specific immunity** because they are produced in response to particular antigens (an antibody produced against *Salmonella* cell walls will not react with *Proteus* cell walls).

The exercises include examples of both natural immunity (exercises 27 and 28) and acquired immunity (exercises 29–32). In exercise 27, you study human blood cells and learn how to determine which ones are phagocytic. In exercise 28, you determine the bacteriostatic activity of the enzyme lysozyme, which occurs naturally in phagocytic white blood cells, saliva, nasal secretions, and tears. Lysozyme is able to digest the cell walls of many bacteria.

The acquired immunity exercises demonstrate examples of antigen-antibody reactions, which vary from one another depending on the nature of the antigen. If the antigen is particulate (as in cells or insoluble substances such as **cardiolipin**) an **agglutination,** or clumping, **reaction** will occur that can be observed visually. For nonparticulate antigens, a **precipitin reaction** occurs that can often be observed visually without performing additional visualization procedures (for example, in exercise 32 the use of an agar immunodiffusion for *Candida* identification, in which precipitin lines form in the agar at the site of the antigen-antibody reaction).

Exercises 30 and 31 represent the current state-of-the-art techniques for demonstrating particulate antigen-antibody reactions. One such reaction, latex agglutination, is commonly used for *S. aureus* detection. In exercise 30 you use it for Lancefield grouping of pathogenic streptococci. For this test, serum antibodies are absorbed on the surface of latex beads. When the specific antibody reacts with the *Streptococcus* antigen in question, latex particle aggregation becomes large enough to be seen visually. The other particulate antigen-antibody reaction, ELISA (enzyme-linked immunoassay) is described in exercise 31, where you use it for *Streptococcus* identification. It is also widely employed in other areas such as plant and animal virus identification and for detecting the presence of HIV antibodies.

Some of our cells, although they are part and parcel of us, have not even fixed coherence with our 'rest'. Such cells are called 'free'. . . . The cells of our blood are as free as fish in a stream. Some of them resemble in structure and ways so closely the little free swimming amoeba of the pond as to be called amoeboid. The pond amoeba crawls about, catches and digests particles picked up in the pond. So the amoeboid cells inhabiting my blood and lymph crawl about over and through the membranes limiting the fluid channels in the body. They catch and digest particles. Should I get a wound they contribute to its healing. They give it a chance to mend, by eating and digesting bacteria which poison it and by feeding on the dead cells which the wound injury has killed. They are themselves unit lives and yet in respect to my life as a whole, they are components in that corporate life.

Sherrington, Man or His Nature

Blood Typing
A B O — antigen

← antigens

Rh(D)

1650

150

lymphocytes : |||||||||||||||||||||||||||||||||

neutrophils : |||||||||||||||? ? ||||

monocytes : |||||| |i

basophils :

eosinophils :

White Blood Cells (WBC)   Red Blood Cells (RBC)

Granulocytes (granule) Neutrophils — light purple 37-80%   non-nucleated biconcave disks

Agranulocytes   0-25% Basophils — Blue smaller — pinkish color

lymphocytes 10-50%   0-7% Eosinophils — Red

monocytes 0-15%

platelets   ◯ purple   neutrophils (usually 7 different lobes)
                  pink

CBC — complete blood count — % of different blood cells (machine)

normal ranges   WBC 4,600 ~ 10,200 /mL

RBC 4,000 — 6,000 /mL   ◯ dark purple
                             ← basophil
platelets 140,000 — 400,000/mL

light blue   ◯ light purple
dark blue lymphocytes   platelets

EXERCISE 27

Differential White Blood Cell Stains

Getting Started

This exercise deals with the cellular forms of the immune system, specifically the white blood cells. For the most part, they can be distinguished from one another using a blood smear stained with a differential stain such as Wright's stain. This stain uses a combination of an acid stain such as eosin and a basic stain such as methylene blue. They are contained in an alcoholic solvent such as methyl alcohol for fixation of the stains to the cell constituents, particularly since the basophilic granules are known to be water soluble. With this stain, a blood smear shows a range in color from the bright red of acid material to the deep blue of basic cell material. In between are neutral materials that exhibit a lilac color. There are also other color combinations depending upon the pH of the various cell constituents.

The two main groups of WBCs are the granulocytes (cytoplasm containing granules) and the agranulocytes (clear cytoplasm). The granulocytes are highly phagocytic and contain a complex, segmented nucleus. The agranulocytes are relatively inactive and have a simple nucleus or kidney-shaped nucleus. Common cell types found in the granulocytes are **neutrophils, eosinophils,** and **basophils.** The basic agranulocyte cell types are the **lymphocytes** and **monocytes.** Another white cell type found in blood is the platelet (very small, multinucleate, irregular pinched off parts of a **megakaryocyte**). Platelets aid in the prevention of bleeding. The appearance of these cell types in blood stained with a differential stain are illustrated in figure 27.1.

Differential blood stains are important in disease diagnosis, since certain WBCs either increase or decrease in number, depending on the disease. In making such judgments, it is important to know the appearance of normal blood (color plate 23). The microscopic field shown includes mostly RBCs with a few neutrophils, only one lymphocyte, and some platelets.

Figure 27.1 Blood cell types present in human peripheral blood. The granular leukocyte names find their origin from the color reaction produced by the granules after staining with acidic and basic components of the staining solution. Neutrophil = neutral or lavender-colored granules; basophil = basic or blue color; and eosinophil = acid or red color.

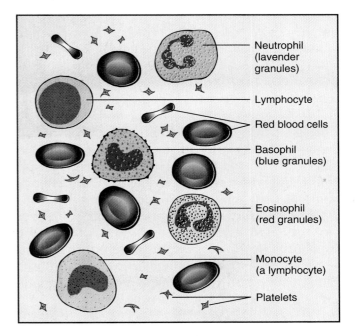

A quantitative description of the various cell types found in normal blood is shown in table 27.1. The red blood cells (RBCs), which are also called erythrocytes, make up the largest cell population. RBCs constitute an offensive weapon because they transport oxygen to various body parts as well as break down carbon dioxide to a less toxic form. The red blood cells in humans and all other mammals (except members of the family Camelidae, such as the camel) are biconcave, circular discs without nuclei.

RBCs are produced in the red bone marrow of certain bones. As they develop, they produce massive quantities of hemoglobin, the oxygen transporting

Table 27.1   Cellular Description of Normal Blood*

Total Counts	Differential WBC Counts
RBC 5,200,000/ml	Neutrophils 64%
WBC 7,200/ml	Leukocytes 33%
Platelets 350,000/ml	Monocytes 2%
Basophils 50/ml	Eosinophils 1%

*From Kracke (see References)

pigment that contains iron, which in the oxygenated form gives blood its red color. Worn out RBCs are broken down at the rate of 2 million cells per second in the liver and spleen by phagocytic white blood cells (WBCs). Some of the components of the RBCs are then recycled in order for the body to maintain a constant number of RBCs in the blood.

The WBCs, or leukocytes, represent approximately 1/800 of the total blood cells (see table 27.1). They are defensive cells, specialized in defending the body against infection by microorganisms and other foreign invaders. Many of the leukocytes are **amoeboid,** capable of moving independently through the bloodstream. They also move out into the tissues where they repel infection and remove damaged body cells from bruised or torn tissue.

In this exercise, you have an opportunity to prepare, stain differentially, and observe some human blood slides. For this purpose, the blood used can be outdated whole blood obtained from a blood bank. If this is not available, commercially prepared blood slides are available.

Definitions

Amoeboid.   To make movements or changes in shape by means of protoplasmic flow.

Basophil.   A granulocyte in which the cytoplasmic granules stain dark purplish blue with methylene blue, a blue basophilic-type dye found in Wright's stain.

Eosinophil.   A granulocyte in which the cytoplasmic granules stain red with eosin, a red acidophilic type dye found in Wright's stain.

Lymphocyte.   A colorless agranulocyte produced in lymphoid tissue. It has a single nucleus with very little cytoplasm.

Megakaryocyte.   A large cell with a lobulated nucleus that is found in bone marrow, and is the cell from which platelets originate.

Monocyte.   A large agranulocyte normally found in the lymph nodes, spleen, bone marrow, and loose connective tissue. It is phagocytic with sluggish movements. When stained with Wright's stain, it is difficult to differentiate from a junior neutrophil.

Neutrophil.   A mature granulocyte present in peripheral circulation. The cytoplasmic granules stain poorly or not at all with Wright's stain. The nuclei of most neutrophils are large, contain several lobes, and are described as polymorphonuclear (PMN) leukocytes.

Plasma.   The fluid portion of the blood before clotting.

Objectives

1. To provide historical and background information on blood and some of its microscopic cell types, their origin, morphology, number, and role in fighting disease.
2. To prepare two stained blood slides: the first for use in observing the cellular appearance of normal blood, and the second slide for use in determining a differential WBC count.

References

Johnston, R. "Monocytes and macrophages." *New England Journal of Medicine* 318 (no. 12): 747–752, 1988.

Kracke, Roy R. *Diseases of the blood and atlas of hematology,* 2d ed. Philadelphia: J. B. Lippincott Co., 1941. Excellent sourcebook, beautifully illustrated, containing clinical and hematologic descriptions of the blood diseases, and a section on technique and terminology. Unfortunately, this is the latest edition.

Lechevalier, Hubert A., and Solotorovsky, Morris. *Three centuries of microbiology.* New York: McGraw-Hill Book Co., 1965. Contains historical information on the immune response.

Metschnikoff, Élie. 1884. "A disease of *Daphnia* caused by a yeast." In *Milestones in Microbiology*, translated by Thomas Brock, American Society for Microbiology, Washington, D.C., 1961. Contains historical information on the immune response.

Nester et al. *Microbiology: A human perspective*. Chapters 15, 346–350; 16, 362; 17, 385; and 28, 625–626.

Materials

Either outdated blood bank whole blood or prepared commercial unstained or stained human peripheral blood smears. For student use, used, outdated whole blood should be dispensed with either a Pasteur pipet or a dropping bottle capable of dispensing a small drop.

Note: In the event of spilled blood, use disposable gloves and towels to remove blood. Then disinfect the area with a germicide such as hypochlorite bleach diluted approximately 1:20 with water. See Nester et al. p. 211 and Notes to the Instructor for additional comments.

For use with whole blood:

New microscope slides, 3

Pasteur pipet for dispensing blood on slides

Hazardous waste container for pipets and slides

For use with whole blood and unstained prepared slides:

Wright's stain, dropping bottle, 1 per 2 students

Phosphate buffer, pH 6.8, dropping bottle, 1 per 2 students

Wash bottle containing distilled water, 1 per 2 students

A Coplin jar with 95% ethanol

Staining rack

Colored pencils for drawings: pink, blue, purple, or lavender

Procedure

Safety Precautions: When using whole blood, be careful not to dispense on surfaces other than slides. Wipe up any spilled blood immediately and disinfect area since blood residues promote the growth of microbial contaminants. Wash and rinse hands immediately after preparing slides for staining.

1. Prepare three clean microscope slides free of oil and dust particles as follows:
   a. Wash slides with a detergent solution, rinse thoroughly.
   b. Immerse slides in a jar of 95% alcohol.
   c. Air dry and polish with lens paper.
2. Place a drop of blood on one end of a clean slide (figure 27.2a). Repeat with a second clean slide.
3. Spread the drop of blood on the slide as follows:
   a. Place the slide on your laboratory bench top. With your thumb and middle finger, firmly hold the sides of the slide on the end where the drop of blood is located.
   b. With your other hand, place the narrow edge of a clean slide approximately ¹/₂″ in *front* of the drop at an angle of about 30° (figure 27.2b).
   c. Carefully push the spreader slide *back* until it comes in contact with the drop, at which point the drop will spread outward to both edges of the slide (figure 27.2c).
   d. Immediately with a firm steady movement *push* the blood slowly toward the opposite end of the slide (figure 27.2d).

   Note: Use of the above procedural restraints (a small drop, a small spreader slide angle, and a *slow*, steady spreader slide movement) should provide a thin film for study of red cells. A good smear has the following characteristics: smooth, without serrations; even edges; and distributed uniformly over the middle ²/₃ of the slide.

   e. Allow slide to air dry for 5 minutes. Do not blot.
   f. For the second slide, prepare a thicker film by using a larger spreader slide angle (45°), and by pushing the blood more rapidly to the opposite end of the slide. The second slide is best for determining the differential white blood cell count.

   Note: The unused end of the first spreader slide can be used to prepare the second slide. Discard used spreader slide in the hazardous waste container.

Figure 27.2 (a–d) Method for preparing a blood smear.

(a) A small drop of blood is placed about 3/4 inch away from one end of slide. The drop should not exceed 1/8" diameter.

(b) The spreader slide is moved in the direction of the arrow (*back*).

(c) Drop of blood spreads outward to both edges of slide.

Thin smear

(d) The spreader slide is pushed along the slide, dragging the blood over its surface.

4. Stain the blood smears with Wright's stain as follows:
   a. Suspend the slides such that they lie *flat* on the staining rack supports.
   b. Flood or add 15 drops of Wright's stain to each blood smear. Stain for 3 to 4 minutes.
   c. Without removing stain, add an equal volume of phosphate buffer. Blow gently on the side of each slide to help ensure mixing of stain and buffer solutions.
   d. Let stand until a green, metallic scum forms on the surface of the slide (usually within 2 to 4 minutes).
   e. Wash off the stain with water. Begin washing while the stain is on the slide in order to prevent precipitation of dye on it. The initial purple appearance on the slide should be washed until it is a lavender-pink.
   f. Wipe off excess stain from the back of the slide and allow it to stand on end to dry (which is preferable to drying between bibulous paper).

5. Examine stained blood smears:
   a. Make an initial examination of the first blood smear with the low power objective to find the most suitable areas for viewing with the oil-immersion objective.
   b. Next using the oil-immersion lens, make a study of the various WBC types present: basophils, eosinophils, lymphocytes, monocytes, neutrophils, and platelets. For help in this study, consult color plate 23 and the Definitions section describing their staining characteristics with Wright's stain. Prepare color drawings of your findings on the Laboratory Report sheet.
   c. Conduct a differential white blood cell count using the second blood smear. For normal blood with a leukocyte count of 5,000 to 10,000 WBCs/ml, one would classify 100 leukocytes. In order to do this, you may have to examine the number and kinds of WBCs present in as many as twenty microscopic fields. Record your findings in table 27.2 of the Laboratory Report and calculate the percentage of each WBC type.

EXODUS

Laboratory Report: Differential White Blood Cell Stains

Results

1. Color drawings of RBCs and various WBCs found in blood smears stained with Wright's stain.

    RBCs   Neutrophils   Eosinophils

    Basophils   Monocytes   Lymphocytes

2. Differential WBC count. In table 27.2, record the kinds of leukocytes found as you examine each microscopic field. After counting 100 WBCs, calculate their percentages from the totals found for each type. Also record the number of microscopic fields examined to find 100 WBCs: _____.

Table 27.2   Kinds and Percentages of WBCs Found in Blood Smear

Neutrophils	Eosinophils	Basophils	Lymphocytes	Monocytes
Total				
Percent				

Questions

1. What problems if any did you find in preparing and staining your blood smears? Indicate any differences noted between thin and thick smears.

2. Were your blood stains satisfactory? Did the stained cells resemble those in figure 27.1 and color plate 23? Were they better?

3. Did your differential white blood cell count percentages compare with the percentages in normal blood (table 27.1)? If not, give an explanation.

4. Were there any WBC types that you did not find in your blood smear? If so, which one(s)? Why did you not find them?

5. Matching (you may wish to consult your text)

   a. Neutrophils   ____ Involved in antibody production

   b. Basophils   ____ A minor phagocytic cell

   c. Monocytes   ____ Increased number in parasitic infections

   d. Eosinophils   ____ Largest WBC

   e. Lymphocytes   ____ Inflammatory WBC

EXERCISE

Lysozyme, an Enzymatic Form of Natural Resistance

Getting Started

A number of antimicrobial chemicals have been isolated from animal cells and body fluids. Among these are two important proteins: **lysozyme** and **complement,** both of which are examples of **natural immunity.** Complement is necessary for certain antigen-antibody reactions in which it becomes fixed to the antigen-antibody complex. For example, the Wassermann complement fixation test is used for syphilis diagnosis. It was developed by Bordet and Gengou in 1901. Through textbook reading, you should become familiar with the principles of this classic syphilis test. (See Nester et al., page 398.)

Lysozyme is a **proteolytic enzyme** found in saliva and nasal secretions. It is also found in phagocytic WBCs (see exercise 27), where it functions as a biocide, and in egg white. Enzymatically, it is able to degrade the peptidoglycan layer of the bacterial cell wall, thereby weakening and eventually destroying it. It is particularly active against gram-positive bacteria that have an exposed peptidoglycan layer. Experimentally, when bacterial cells are treated with lysozyme in a solution of higher osmotic pressure, such as 0.6 M sucrose, the cell wall dissolves but cell lysis is inhibited (see exercise 13).

Chemically, lysozyme is a **globulin protein** (N-acetyl-muramide hydrolase) that was discovered by Alexander Fleming in 1922. It functions by dissolving the **peptidoglycan layer** of the bacterial cell wall (see Nester et al., chapter 3, pages 64–65 for more details). It also acts on **chitin,** which is the principal component in the exoskeleton of molds, yeasts, invertebrates, and arthropods. Another source of lysozyme is the tail section of the bacteriophage. The lysozyme not only helps with phage entry through the bacterial cell wall, but also later phage stages direct the host cell to produce more lysozyme for use in facilitating release of new phage particles.

In this exercise, you have an opportunity to assay the antimicrobial activity of lysozyme collected from tears and egg white, and to compare their antimicrobial activity with a commercial lysozyme preparation of known activity.

Definitions

Chitin. A polysaccharide structurally similar to cellulose.

Complement. A system of at least 26 serum proteins that act in sequence, producing certain biological effects concerned with inflammation and the immune response.

Globulin protein. A class of simple proteins characterized by their almost complete insolubility in water, solubility in dilute salt solutions, and coagulability by heat. They occur widely in plant and animal tissues as blood plasma or serum.

Lysozyme. An enzyme able to attack and destroy bacterial cell walls. It occurs naturally in tears, saliva, phagocytic WBCs, and egg white.

Natural immunity. An immunity to infectious disease in a species occurring as a part of its natural biologic makeup.

Peptidoglycan layer. The rigid backbone of the bacterial cell wall, composed of repeating subunits of N-acetylmuramic acid and N-acetylglucosamine and other amino acids.

Proteolytic enzyme. An enzyme able to hydrolyze proteins or peptides with the resulting formation of simpler, more soluble products such as amino acids.

Objectives

1. To become more familiar with the process of natural immunity, and how chemicals such as complement and lysozyme function in immunity.

2. To learn how to determine the antimicrobial activity of various natural lysozyme preparations.

References

Fleming, A. 1922. Proc. of the Royal Soc. of London, Ser. B, 93, 306.

Nester et al. *Microbiology: A human perspective.* Chapters 3, 64–65; 15, 342–343, 350–351; and 33, 720–721.

Osserman, et al. eds. *Lysozyme.* New York: Academic Press, 1974.

Materials

Per pair of students

Nutrient broth cultures (37° C, 24 hour) of *Staphylococcus epidermis* and *E. coli*

Sabouraud's dextrose broth culture (25° C, 24 hour) of *Saccharomyces cerevisiae*

Melted nutrient agar deeps, 2, and Sabouraud's dextrose agar deep, 1, held in a 48° C water bath

Sterile Petri dishes, 4

Petri dish containing 9 sterile filter paper discs (approximately ½″ diameter)

Petri dish containing 1–2 ml of aseptically prepared raw egg white

Petri dish containing 1–2 ml of Lysozyme chloride (Sigma cat. # L-2879) with activity of approximately 60,000 units per mg of protein. Diluted 1:10 with sterile distilled water

Raw onion

Scalpel or sharp knife for cutting the onion

Mortar and pestle

Pair of tweezers

Sterile 1-ml pipets, 3

Ruler calibrated in mm

Procedure

First Session

1. With a glass marking pencil, divide the underside of 3 Petri dishes in thirds. Label one part T for tears, another part EW for egg white, and the remaining part L for lysozyme.

2. Label each Petri dish with the name of one of the three test organisms. Also include your name and the date on the bottom.

3. Prepare pour plates of *Staphylococcus* and *E. coli* by suspending the broth culture. With a sterile 1-ml pipet aseptically transfer a 1-ml aliquot to the respectively marked Petri dish. Discard pipet in the hazardous waste container. Add the contents of a tube of melted nutrient agar to the dish, replace cover, and rotate the dish 6 to 8 times on the surface of your desk in order to distribute the cells uniformly. Allow agar to harden. Repeat above procedure for the other culture.

4. Repeat step 3 above for *S. cerevisiae* using a melted Sabouraud's agar deep.

5. In order to induce tear secretions, one student should remove the outer skin from an onion. Then cut the onion into small pieces and crush in a mortar with a pestle. The other student should be prepared to collect the secreted tears in a sterile Petri dish—0.5 to 1 ml is sufficient.

6. Alcohol sterilize tweezers and air cool. Aseptically remove a filter paper disc and dip it into the tears. Transfer the moistened disc to the center of the area marked T on the *S. epidermis* agar plate.

7. Repeat step 6 by transferring moistened tear discs to Petri dishes containing the remaining two test organisms.

8. Repeat steps 6 and 7 with the egg white preparation.

9. Repeat steps 6 and 7 with the lysozyme preparation.

10. Incubate the Petri dishes containing *Staphylococcus* and *E. coli* at 37° C for 24 hours and the Petri dish containing *S. cerevisiae* at 25° C for 48 hours.

Second Session

1. Observe the Petri dishes for zones of inhibition around the filter paper discs. With a ruler, determine their diameter in mm and record your results in table 28.1 of the Laboratory Report.

EXERCISE

Laboratory Report: Lysozyme, an Enzymatic
Form of Natural Resistance

Results

1. Complete table 28.1 (see instructions in Second Session, Procedure step 1).

Table 28.1   Antimicrobial Activity of Various Lysozyme Extracts

Test Organism	Diameter of Inhibition Zone (mm)		
	Tears	Egg White	Lysozyme*
S. epidermis			
E. coli			
S. cerevisiae			

*1:10 dilution

2. Which of the three preparations was the most active? _____ Least active?
    _____ Consider the lysozyme *dilution* when preparing your answer.
3. Which organism(s) were not inhibited by lysozyme?

4. Knowing that the above lysozyme preparation has an activity of 60,000 units per mg of protein, calculate the lysozyme activity in similar units for tears and egg white against *S. epidermis* and *E. coli*. Some of your results from table 28.1 are necessary for making this calculation. Record your findings in table 28.2.

Table 28.2   Units of Lysozyme Activity for Egg White and Tears

Test Organism	Units of Lysozyme Activity/Mg of Protein	
	Egg White	Tears
S. epidermis		
E. coli		

Questions

1. What similarity do lysozyme and penicillin have in their mode of antimicrobial action? How do they differ?

2. Why must tweezers be cool before dipping the filter paper discs in the lysozyme solutions?

3. What large groups of microorganisms are susceptible to lysozyme? Resistant to lysozyme?

4. Procedurally, what additional important step would be necessary to evaluate the lysozyme activity of nasal secretions? Why?

5. Why are most gram-negative bacteria not lysed by lysozyme, yet they have a peptidoglycan cell wall structure similar to that of the gram-positive bacteria?

NOTES:

lysozyme — an enzyme that is one of our bodies natural immunity factors.

found in saliva, mucous, tears

proteolytic enzyme — breaks down proteins xchitin(yeasts)

especially effective against gram +

NA E. coli   g- rod
NA S. epidermitis g+ cocci
Sab. S. cerevisine — yeast

1) saliva
2) lysozyme
3) egg white

EXERCISE

Traditional Agglutination Reactions Employing Microbial and Nonmicrobial Antigens

Getting Started

An **agglutination** reaction is an antigen-antibody reaction in which the antigen consists of particulate matter such as cells or synthetic material and the antibodies are described as **agglutinins.** The usefulness of this simple procedure—mixing antigen particles with antibody—gave rise to the era of **serodiagnosis,** advanced understanding of the role microorganisms play in causing disease. It eventually led to the discovery of the ABO blood groups.

Agglutination occurs in two steps: the specific combination of antigen and antibody, which is then followed by the visible aggregation of the particles. Factors such as the charge of the antigen-antibody particles, buffering, and viscosity of the test medium play a role in proper agglutination of antigen particles by antibodies. A disadvantage of the agglutination phenomenon is that the reaction is semiquantitative (only accurate to a fourfold difference in antibody **titer**). However, the facts that numerous systems lend themselves to agglutination reactions, the basic simplicity of agglutination systems developed to date, and the high sensitivity of agglutination-based reactions encourage wide use of such tests.

Use of Microbial Antigens for Diagnosis of Infectious Diseases

Some of the most useful agglutination tests are used for infectious disease diagnosis, such as for salmonellosis and rickettsiosis. Early identification of initially high or rising agglutinin titers to these organisms offers a powerful laboratory adjunct to clinical diagnosis.

For example, the Widal test, which was devised in 1954, is used to identify serum antibodies against various *Salmonella* species antigens, which vary in composition from body parts (**somatic** and capsular antigens) to flagellar antigens. A titer of somatic and flagellar antibodies equal to $1/_{80}$ or greater suggests an active infection.

Table 29.1 Weil-Felix Reactions in Rickettsioses*

Disease	OX-19	OX-2	OX-K
Epidemic typhus	++++	+	0
Murine typhus	++++	+	0
Brill-Zinsser disease	variable	variable	0
Spotted fever[a]	++++	+	0
	+	+++	0
Rickettsial pox	0	0	0
Scrub typhus	0	0	+++
Q fever	0	0	0
Trench fever	0	0	0

[a]Spotted fever immune sera can agglutinate either OX-19 or OX-2, or both.
*From the *Manual of Clinical Immunology,* p. 709, 1980.
National Committee for Clinical Laboratory Standards. *Performance Standards for Antimicrobial Disk Susceptibility Tests.* Fifth Edition; Approved Standard. Copyright © 1993 The National Committee for Clinical Laboratory Standards. By permission.

Another useful test is the Weil-Felix test, which will be evaluated in this exercise. It uses *Proteus* antigens to detect **cross-reacting** rickettsial serum antibodies. Rickettsial diseases diagnosed most frequently in the United States include murine typhus fever, Q fever, and Rocky Mountain spotted fever. The latter is caused by *Rickettsia rickettsii,* which is transmitted to humans by a tick **vector.** Thus, the Weil-Felix reaction is useful in screening clinically suspect patients for high titers of *Proteus* agglutinins for presumptive diagnosis of Rocky Mountain spotted fever and murine typhus fever (whose vector is infected flea feces). *Proteus* antigens are used because the two rickettsia share some minor antigens with these bacteria.

Table 29.1 shows that *Proteus vulgaris* strains OX-19 and OX-2 are agglutinated by sera of individuals infected with most members of the spotted fever group. Rickettsial antibodies can reach detectable levels within one week after the onset of symptoms, and usually reach a maximum titer within a few months.

Although the Weil-Felix test is the least sensitive of immunological tests currently available for use in rickettsial disease diagnosis, it is used in this exercise to illustrate a classical agglutination reaction. It is also the least specific test since the *Proteus* organisms that are used are also agglutinated by antibodies occurring as a result of *Proteus* urinary tract infections.

Antigens available for the Weil-Felix test are suspensions of whole, unflagellated *P. vulgaris* strains OX-19 and OX-2, and *P. mirabilis* strain OX-K. Motility of the organisms should be checked with a hanging drop culture since flagellar H antigens, if present, prevent agglutination of the somatic O antigens by rickettsial antibodies. Controls for the W-F test should include a positive human antirickettsial serum and a normal human serum (negative control).

The W-F test can be performed by both rapid slide and tube agglutination methods. The former is a qualitative method and the latter a quantitative method. When conducting the test, keep in mind that it is not specific for rickettsial infections. Agglutination may occur from a previous *Proteus* infection and also from other cross-reacting organisms. However, if properly performed, the test remains a useful tool in the diagnosis of rickettsial disease, due to wide availability of antigens.

Use of a Nonmicrobial Antigen for Initial Syphilis Diagnosis

Tests for syphilis diagnosis involve two generations of antigens: the first was **cardiolipin** (derived from beef heart) and the second was the infectious disease organism *Treponema pallidum*. It is paradoxical that the use of cardiolipin still remains superior to the use of *T. pallidum* even though it is biologically nonspecific. In this test, cardiolipin antigens are used to detect antilipid antibodies, traditionally termed **reagin,** in the patient's serum. It is not known whether these antibodies are invoked by lipid antigen present in *T. pallidum* or by the host-parasite reaction.

The nontreponemal tests most frequently used in the United States today are the Venereal Disease Research Laboratory (VDRL) test and the rapid plasma reagin (RPR) 18 mm circle card test. Both are simple, rapid, reproducible, and inexpensive; both can be used either qualitatively or quantitatively.

The RPR card test has various advantages over the VDRL test: it is available as a kit test containing all needed reagents and controls including a prepared antigen suspension; unheated serum is used; the reaction is read macroscopically; and most materials are throwaways.

The unheated serum reagin (USR) test evaluated in this exercise is a cross between the VDRL and RPR syphilis diagnosis tests. The USR test, like the RPR test, uses unheated serum and cannot be used to examine CSF. However, unlike the RPR test and like the VDRL test, agglutination must be observed microscopically rather than macroscopically. All three tests are alike in that they use the nonmicrobial antigen cardiolipin.

The nontreponemal tests are of greatest value when used as screening procedures and for evaluating patient response to syphilis therapy. Treponemal tests are used primarily to determine whether a reactive nontreponemal test is due to syphilis or some other condition. The treponemal tests can also be used to detect syphilis in patients with negative nontreponemal tests but with clinical evidence of syphilis. False-positive treponemal tests should be further evaluated. Unlike the nontreponemal test, the treponemal tests do not indicate the patient's response to treatment, and quantitative tests are of no value in diagnosis or prognosis. Finally, there are no standard procedures for testing CSF with treponemal tests.

Although the immunological response to infection with *T. pallidum* is complex and poorly understood, serological tests are frequently the only basis for syphilis diagnosis and for evaluating patient response to treatment (Coffey and Bradford, 1980). Nontreponemal tests usually become reactive 4–6 weeks after infection or 1–2 weeks after the first visible chancre appears. The specificity of nontreponemal tests is lacking since they can be reactive in a variety of other conditions. Treponemal tests, although specific, vary in their ability to react in early syphilis. Once reactive, all treponemal tests tend to remain so for years. None of the treponemal tests distinguish between syphilis and other treponemal infections, such as bejel, pinta, and yaws.

Definitions

Agglutination. The aggregation by antibodies (agglutinins) of cells or synthetic particles (agglutinogens) into visible clumps that settle to the bottom.

Agglutinin. A specific antibody capable of affecting the agglutination of the agglutinogen that stimulated its production.

Cardiolipin. A phosphatide obtained from beef heart and used as an antigen in diagnostic tests for syphilis.

Cross-reacting antibodies. The immunological phenomenon wherein one antigen reacts with antibodies that were developed against another antigen.

Reagin. A substance in the blood of people with syphilis that sometimes functions as an antibody against a syphilis antigen.

Serodiagnosis. Diagnosis of disease by the use of serum as in the USR test for syphilis and the Weil-Felix test for rickettsial diseases.

Somatic antigen. An antigen from the body of the bacterial cell.

Titer. A measure of the serum antibody level. Usually measured as the highest dilution of serum that will give a positive test for that antibody. The titer is often expressed as the reciprocal of that dilution.

Vector. An agent, often an insect, that transmits an infectious disease from one host to another host. For example, fleas transmit bubonic plague from rats to humans.

Objectives

1. To provide background information about agglutination tests using both microbial and nonmicrobial antigens.

2. To describe some of the advantages and disadvantages of these tests when used in diagnosis and for evaluating patient response to therapy. Diseases discussed are Rocky Mountain spotted fever (a rickettsial disease) and syphilis (a treponemal disease).

3. To evaluate in the laboratory a conventional agglutination method, the Weil-Felix test, which uses a microbial *Proteus* antigen for presumptive diagnosis of rickettsial diseases. The test is evaluated qualitatively by the rapid slide method and quantitatively by the tube dilution method.

4. To evaluate an agglutination reaction with the nonmicrobial antigen cardiolipin, for use in presumptive syphilis diagnosis, the USR (unheated serum reagin) test.

References

Coffey, E., and Bradford, L. "Serodiagnosis of Syphilis," 530–541, in the *Manual of clinical immunology* (see below).

D'Angelo, L. J., Winkler, W. G. and Bregman, D. J. 1978 "Rocky Mountain spotted fever in the United States, 1975–77." *Journal of Infectious Diseases* 138:273–276.

Manual of Clinical Immunology, 2d ed. Eds. Noel R. Rose and Herman Friedman, American Society for Microbiology, Washington, D.C., 1980. Contains a wealth of information presented by various authors clearly and concisely. A must for any instructor or student desiring to obtain a solid introduction to clinical immunology. See also 4th ed., 1992.

Manual of Tests for Syphilis. Venereal Disease Program, 1969. Centers for Disease Control, Atlanta.

Nester et al. *Microbiology: A human perspective*. Chapters 10, 228; 11, 239, 245–246; 16, 357–361; 17, 394–398; 19, 421, 431–433; and 24, 556–557.

Materials

Use of *Proteus* Antigens to Detect Rickettsial Antibodies.

For use with rapid slide and tube test:

Proteus OX-2 or OX-19 antigen and antiserum, for source see Notes to Instructor

Suitable light source for observing agglutination, such as a goose-neck, fluorescent, or fiber optic lamp

A rubber bulb for pipetting serum and antiserum

Isotonic saline (0.85 g NaCl/100 ml distilled water)

For use with rapid slide test:

Clean microscope slides, 3

Applicator sticks or toothpicks

Dropper delivering approximately 0.03 ml

Serological pipets, 0.2 ml, cotton plugged, 2

For use with tube test:

Serology test tubes, approximately 10 × 100 mm, 10

Test-tube rack for holding serology test tubes

Sterile 5-ml serological pipets, cotton plugged, 2

Sterile 1-ml serological pipets, cotton plugged, 2

Water bath, 37° C

Use of a Nonmicrobial Antigen (Cardiolipin) for Syphilis Diagnosis.

USR antigen and USR test control serum set, for source see Notes to Instructor

Negative control serum

Hypodermic needle without bevel, 18 gauge adjusted to drop $^1/_{45}$ ml per drop

Syringe, Luer-type, 1–2 ml

Absolute alcohol and acetone for rinsing syringe with needle

Clean microscope slides, 2

Serological pipets, 0.2 ml, cotton plugged, 3

Figure 29.1 Rotation method used to initiate agglutination of antigen-antibody mixture.

Procedure

The Rapid Slide Agglutination Test: A Qualitative Screening Test Employing a Microbial (*Proteus*) Antigen

1. With a glass-marking pencil, mark three *clean* microscope slides with two 16-mm ($^5/_8''$) circles per slide.
2. Using a 0.2-ml pipet with a rubber bulb, pipette the following amounts of *Proteus* antiserum into the first five circles: 0.08, 0.04, 0.02, 0.01, and 0.005 ml. Discard used pipet in hazardous waste container. To the sixth circle add 0.08 ml of 0.85% saline using a fresh pipet (negative control).
3. To each circle, add one drop of *Proteus* antigen with a dropper.

   Note: Shake the antigen well before using.

4. Mix each antiserum-antigen composite with an applicator stick or toothpick. Start with the 0.005-ml serum dilution and work back to the 0.08-ml dilution. Discard mixing tool in the hazardous waste container. The final dilutions correspond approximately with the macroscopic tube test dilutions, which are 1:20, 1:40, 1:80, 1:160, and 1:320, respectively.
5. Hold the slide in both hands and gently rotate 15–20 times (figure 29.1).

Figure 29.2 Serial tube dilution test protocol employing doubling dilutions.

	#1	#2	#3	#4	#5	#6	#7	#8	#9	#10
Amount of saline (ml)	0.9	0.5	0.5	0.5	0.5	0.5	0.5	0.5	0.5	0.5
Amount of serum (ml)	0.1	0.5 #1	0.5 #2	0.5 #3	0.5 #4	0.5 #5	0.5 #6	0.5 #7	0.5 #8	
Initial dilution	1/10	1/20	1/40	1/80	1/160	1/320	1/640	1/1280	1/2560	Control
Amount of antigen (ml)	0.5	0.5	0.5	0.5	0.5	0.5	0.5	0.5	0.5	0.5
Final dilution	1/20	1/40	1/80	1/160	1/320	1/640	1/1280	1/2560	1/5120	Control

6. Observe for macroscopic agglutination (clumping) using a suitable light source.

   Note: Make observations within 1 minute, since later reactions may be due to drying of reactants on slide.

7. Record the amount of clumping for the various dilutions in table 29.2 of the Laboratory Report as follows:

   – no agglutination
   + trace agglutination
   1+ approximately 25% cell clumping
   2+ approximately 50% cell clumping
   3+ approximately 75% cell clumping
   4+ complete agglutination

The Tube Dilution Agglutination Test: A Quantitative Test Employing a Microbial (*Proteus*) Antigen

Note: Consider use as a demonstration exercise (see Appendix 5, Notes to Instructor).

1. Prepare serial dilutions containing 0.5 ml of positive control antiserum in serology test tubes as follows (figure 29.2):
   a. Place ten serology test tubes in a test-tube rack.
   b. With a 1-ml serological pipet, transfer 0.9 ml of 0.85% saline into the first tube and 0.5 ml into each of the remaining tubes. Discard pipet.
   c. With a fresh 1-ml serological pipet containing a rubber bulb, add 0.1 ml of antiserum to the first tube. Mix well by drawing in and out of pipet about 3 times.
   d. With the same pipet, transfer 0.5 ml from tube 1 to tube 2. Mix and transfer 0.5 ml of tube 2 to tube 3, mix; continue above dilution process through tube 9, discarding 0.5 ml from tube 9 along with pipet into hazardous waste container.

   Note: Tube 10, which does not contain serum, serves as an antigen control. Also note that all tubes now contain 0.5 ml. Also, the serum dilution in the first tube is 1:10 because 0.1 ml serum was added to 0.9 ml saline, and the serum dilution in tube 2 is 1:20 because 0.5 ml of the 1:10 dilution was added to an equal volume of saline. The same principle holds for tube 3, which is 1:40 or double that of tube 2. These are called doubling dilutions.

2. With a 1-ml serological pipet containing a rubber bulb, add 0.5 ml of well suspended *Proteus* antigen to each of the ten tubes. Discard pipet in hazardous waste container.

Figure 29.3 Examination of tubes in an agglutination test. (*a*) Negative control and (*b*) positive agglutination result.

Eye point

(a)   (b)

3. Gently shake the rack to mix antigen and antiserum. The resultant dilutions are 1:20 through 1:5120, respectively.
4. Incubate the rack in a 37° C water bath for 2 hours, followed by overnight incubation in a 2–8° C refrigerator.

Note: It is important to use the recommended time and temperature of incubation, and to make certain that the water bath is in a location free of mechanical vibration.

As you have seen, the above exercise was conducted using a positive control antiserum. In a hospital environment, serum drawn from a patient would constitute the unknown test serum. For greater proficiency in test interpretation, it is also important to include a febrile negative control antiserum. The latter is processed in the same manner as the above positive control antiserum.

5. Examine the tubes *qualitatively* for the type of sediment using a mirror, preferably concave, under a light (figure 29.3).

The negative control tube should show that the antigen has settled out in the bottom of the tube in a small, round disc with smooth edges (figure 29.3*a*). In positive tubes, the cells settle out over a larger area, and the edges are irregular (figure 29.3*b*). Record your results in table 29.3 of the Laboratory Report.

6. Examine the tubes *quantitatively* for agglutination by gently mixing the contents (easily accomplished by flicking the tube back and forth with your index finger), and observing the tube with a good light source against a dark background. Record the amount of agglutination in table 29.4 of the Laboratory Report as follows:

– no agglutination, cells remain in a cloudy suspension
1+ approximately 25% cell clumping, supernatant cloudy
2+ approximately 50% cell clumping, supernatant moderately cloudy
3+ approximately 75% cell clumping, supernatant slightly cloudy
4+ 100% cell clumping, supernatant clear

The Use of a Nonmicrobial Antigen (Cardiolipin) for Initial Syphilis Diagnosis (the USR Test)

Note: The USR test is performed as described in the *USPHS Manual of Tests for Syphilis 1969* and its supplement, January 1982.

1. Thoroughly soak and wash two glass slides in a glassware detergent solution. Then rinse with tap water 3–4 times, followed by a distilled water rinse, finally wiping dry with a clean lint-free cloth. This cleaning procedure will enable the serum to spread evenly within the inner surface of the circle.
2. With a wax marking pencil, inscribe two circles (14-mm diameter) on each of the two slides.

3. The syringe with needle should be washed by prerinsing with tap water, then soak and wash thoroughly in a glassware detergent solution. Follow by rinsing with tap water 6–8 times, then with distilled water, absolute alcohol, and acetone, respectively. Finally, air dry until the acetone odor is gone.

4. The syringe with needle attached should be calibrated for delivery by filling the syringe with antigen suspension. Then hold the syringe in a vertical position and expel the suspension dropwise into the suspension bottle. Count the number of drops delivered per ml of antigen. The needle should deliver 45 drops ± 1.* Adjust drops per ml by either narrowing the open end of the needle to allow more drops per ml to be delivered or opening the end of the needle to allow fewer drops per ml.

5. Using a rubber bulb, pipette 0.05 ml of positive antiserum in the center of one of the circles using a 0.2-ml serological pipet. Spread the serum with the aid of the pipet tip over the area of the circle. Discard the used pipet in a hazardous waste container.

6. To each circle, add 1 drop of cardiolipin antigen using the previously calibrated syringe.

*Note: If unable to properly calibrate the syringe, use a 50 microliter (μl) pipet to deliver 0.05 ml of antiserum and antigen.

7. Rotate the slide for 4 minutes, preferably on a rotating machine at 180 rpm, circumscribing a circle $^3/_4''$ in diameter on a horizontal plane. The slide may also be rotated by hand for 4 minutes (see figure 29.1).

8. Observe the slide *immediately* with the 10× objective of your microscope and record your findings in the Laboratory Report as Reactive, Weakly Reactive, or Nonreactive, as determined by the amount of clumping:

   Reactive = medium and large clumps
   Weakly reactive = small clumps
   Nonreactive = no clumps or slight roughening

9. Repeat the above procedure with the negative control antiserum, and then again with the weakly reactive antiserum.

Note: Each serum reported as positive in the USR qualitative test would normally be subject to further serologic study, including quantitation using a method somewhat like the rapid slide test, and if indicated, to other confirmatory syphilis serology tests, for example, the Fluorescent Treponemal Antibody Absorption (FTA-ABS) or the Hemagglutination Treponemal Test (HATTS). Thus, positive tests with the USR antigen are not conclusive evidence for syphilis. Conversely, a negative USR test by itself does not rule out syphilis diagnosis.

NOTES:

EXERCISE

Laboratory Report: Traditional Agglutination Reactions
Employing Microbial and Nonmicrobial Antigens

Results

1. Rapid slide test with a microbial antigen
   a. Identify the *Proteus* antigen you used: _____
   b. Record the amount of agglutination in table 29.2 below:

Table 29.2   *Proteus* Agglutination with Different Serum Dilutions

Dilution:	1:20	1:40	1:80	1:160	1:320	Saline control
Amount of agglutination:						

   c. Record the highest serum dilution with 2+ agglutination: _____
   d. Record the serum titer as the reciprocal of the highest dilution showing a 2+ reaction: _____

2. Tube dilution test with microbial antigen
   a. Identify the *Proteus* antigen you used: _____
   b. Record the presence (+) or absence (−) of agglutination for the various dilutions in table 29.3 below:

Table 29.3   *Proteus* Agglutination with Different Serum Dilutions

Dilution:	1:20	1:40	1:80	1:160	1:320	1:640	1:1280	1:2560	Control
Agglutination:									

   c. Using protocol outlined in part b of the tube dilution procedure, record the *amount* of agglutination for the various tube dilutions in table 29.4 below:

Table 29.4   Amount of *Proteus* Agglutination with the Tube Test

Dilution:	1:20	1:40	1:80	1:160	1:320	1:640	1:1280	1:2560	Control
Amount of agglutination:									

d. Record the highest serum dilution with 2+ agglutination: _____

e. Record the serum titer as the reciprocal of the highest dilution showing a 2+ reaction: _____

3. Reagin test with a nonmicrobial antigen (cardiolipin)

a. Record your results for the positive, weakly positive, and negative control serums:

Positive Serum: _____   Weakly Positive Serum: _____   Negative Serum: _____

Questions

1. What titers did you find in the bacterial agglutination tests?

_____ Rapid Slide   _____ Tube Dilution

If this were serum from a patient, what would be its significance with respect to the probability of infection?

2. Why is a positive titer with the Weil-Felix test not necessarily conclusive that the patient has a rickettsial infection?

What other tests would be necessary to confirm a rickettsial infection? Consult your text and other references for possible answers, for example the Difco Manual.

3. The rapid slide agglutination test is defined as a qualitative method. Why then are a series of doubling dilutions evaluated for use with this method?

4. Discuss ways in which the rapid slide agglutination test differs from the USR agglutination test. Consider variables such as test antigens and observation techniques.

5. Discuss the pros and cons of using nontreponemal antigens (such as cardiolipin) rather than treponemal antigens for syphilis diagnosis.

NOTES:

EXERCISE

Lancefield Grouping of Pathogenic Streptococci with a Latex Slide Agglutination Test

Table 30.1   Some Characteristics Useful in Identifying the Major Streptococci Associated with Human Disease

Streptococcus Species	Lancefield Group	Hemolysis Type	Human Diseases
S. pyogenes	A	β	Pharyngitis, scarlet fever, wound infections, skin, ear, lungs, other tissues, sequela: glomerulonephritis, rheumatic fever
S. agalactae	B	β	During birth and in infants: may develop pneumonia, septicemia, meningitis
S. pneumoniae	B	α	Bacterial pneumonia, ear infections, meningitis
S. faecalis	D	α or β	Subacute bacterial endocarditis (rare),
S. faecium			urinary tract infections
S. durans			
S. viridans			
(10 species)	Occasional false +	α	Subacute bacterial endocarditis

Getting Started

Special attention has been given to the genus *Streptococcus* because of its involvement in numerous diseases of humans and animals (see exercises 23, 24, and 31). It represents the predominant normal bacterial flora of the human respiratory tract, and is also found in the intestinal and genital tracts. Few other microorganisms of medical importance can elaborate as many **exotoxins** and destructive enzymes, as well as produce serious infections in virtually every tissue as can *Streptococci* (see table 30.1). One of the **sequela** diseases, rheumatic fever, was recently reported on the upsurge in the United States after decades of steady decline (Bisno, 1988). It is caused by a Lancefield Group A streptococcus that initially causes pharyngitis (strep throat).

It is important to differentially diagnose strep throat infections from viral throat infections since both acute rheumatic fever and acute **glomerulonephritis** can occur if prompt appropriate therapy is not initiated (Facklam and Carey, 1985). Rapid immunological slide tests are now available for detecting and differentiating pathogenic streptococci. They have their origin in historical research done by Rebecca Lancefield (1933), who was able to group streptococci serologically

into 18 groups (A through R) on the basis of a group specific carbohydrate antigen present in their cell wall. One exception, is the Group D streptococci in which a noncarbohydrate antigen, teichoic acid, is found in the cytoplasmic membrane. The serological test developed by Lancefield was an overnight capillary precipitin test using extracted cell wall antigens.

Further examination of table 30.1 shows that most human streptococcal diseases are caused by beta (β)-hemolytic streptococci (see exercise 24 for how to determine β-hemolysis) belonging to Lancefield Group A. This species, *S. pyogenes*, in contrast to other pathogenic streptococci, is found in only a small percentage of healthy humans. Other characteristics important for final identification of streptococci include biochemical and morphological tests (see exercises 23 and 24). In conjunction with this exercise, you examine their morphology and hemolysis reactions on blood agar. With one exception, *S. pneumoniae*, all streptococci appear as chains when grown in a broth medium (figure 30.1*a*). On an agar medium they sometimes appear as diplococci (figure 30.1*b*). *S. pneumoniae* is a true diplococcus that can be differentiated from other diplococci by its lancet shape (figure 30.1*c*) and **encapsulation.**

Figure 30.1 Genus *Streptococcus* morphology. (*a*) *S. pyogenes* chains in a broth medium; (*b*) *S. pyogenes*, sometimes diplococci on solid media; (*c*) *S. pneumoniae*, lancet-shaped diplococci in a broth medium. Phase-contrast photomicrographs courtesy of G. E. Roberts

(a)

(b)

(c)

The principle of the latex agglutination test for Lancefield grouping of pathogenic streptococci is that the streptococcus cell wall carbohydrate antigens are allowed to react with specific antibodies coated to the surface of latex beads. The agglutination reaction occurs instantaneously and the latex particle aggregation is large enough to be seen visually with the naked eye, thus eliminating the need for microscopic observation. A necessary preliminary step is the extraction of specific cell wall carbohydrate antigens. The carbohydrate known as "C" carbohydrate is a branched polymer composed of N-acetylglucosamine coupled with certain sugar molecules. It can be extracted from the cell wall with either hot formamide or hot trichloroacetic acid autoclaving, or by enzymatic digestion with lysozyme (see exercise 28) and certain microbial enzymes. The enzymatic process is used in this exercise.

Using the previously discussed identification methods (hemolysis, morphology, and latex agglutination) in this exercise you have an opportunity to differentiate two streptococcus species: a Lancefield Group A species (*S. pyogenes*) and a Lancefield Group B species (*S. pneumoniae*).

Definitions

Beta-hemolysis. A sharply defined, clear, colorless zone of hemolysis surrounding colonies of certain streptococci grown on blood agar plates.

Encapsulation. The surrounding of some bacteria by a protective gelatinous material, which may also relate to their virulence.

Exotoxin. A soluble, poisonous protein that passes into the growth medium during the growth of certain bacteria.

Glomerulonephritis. Inflammation of the kidneys affecting the structure of the renal glomeruli (inflammation of the capillaries caused by toxins produced elsewhere in the body).

Precipitin reaction. The reaction of an antibody with a soluble antigen to form an insoluble, visible antigen-antibody complex.

Sequela. An aftereffect of disease or injury that is often more serious than the initial disease.

Objectives

1. To provide introductory information about the medical importance of the genus *Streptococcus* and how such human pathogens can be identified by a combination of morphological, biochemical (hemolysis), and immunological (latex agglutination) tests.
2. To determine the Lancefield Group, A or B, of two *Streptococcus* species with the latex slide agglutination test.

References

Bisno, Alan, U. of Miami School of Medicine, in the *Seattle Times*, p. A4, January 20, 1988.

Facklam, R. R. 1980. "Isolation and identification of streptococci." HEW Publication, U.S. Dept. of Health, Education, and Welfare, Centers for Disease Control, Atlanta.

Facklam, R. R., and Carey, R. B. 1985. "Streptococci and aerococci," 154–175. In E. H. Lennette, A. Balows, W. J. Hausler, Jr., and H. J. Shadomy (ed.), *Manual of clinical microbiology*, 4th ed. American Society for Microbiology, Washington, D. C.

Lancefield, R. C. 1933. Serological differentiation of human and other groups of hemolytic streptococci. *J. Exp. Med.* 57:571–595. A classical paper worth examining.

Nester et al. *Microbiology: A human perspective.* Chapters 11, 15, 16, 19, 20, 22, 25, 26, 27.

Slifkin, M., and Pouchet-Melvin, G. R. 1980. "Evaluation of three commercially available test products for serogrouping beta-hemolytic streptococci." *Journal of Clinical Microbiology* 11:249–255.

Materials

Fresh (24 hr, 37° C) *unknown* cultures of *S. pyogenes* (Lancefield Group A) and *S. pneumoniae* (Lancefield Group B) on blood agar plates labeled 1 and 2

Fresh TS broth cultures of unknowns 1 and 2 above

Diluted cell wall extraction enzymes (see Appendix 5, Notes to Instructor for dilution procedure), 0.3-ml aliquots contained in 3 serological test tubes

One vial with dropper of a 1% latex bead suspension coated with strep Gr A antibodies, prepared in a glycine buffer (to be shared by class)

A vial (similar to vial above) except coated with Strep Gr B antibodies (to be shared by class)

A vial of polyvalent positive control antigen (an extract of strep Gr A, B, C, F, and G). To be shared by class.

BBL provides above reagents in kit form (see Appendix 5, Notes to Instructor)

Sterile physiological saline (0.85% NaCl)

Toothpicks

37° C water bath

Vortex for mixing tubes

High-intensity incandescent light source

Mechanical rotator for slides (if available)

Ruler calibrated in mm

Procedure

Safety Precautions: All used slides, disposable pipets, and stirring sticks should be disposed of in the hazardous waste container. Any used glassware (blood agar plates, broth cultures, and tubes containing extraction enzyme-unknown bacteria) should be promptly autoclaved. If any culture material is spilled, notify your instructor. All of these precautions are necessary because these bacteria are opportunistic pathogens.

1. Prepare Gram stains of both the broth and agar unknown cultures. Examine with the oil-immersion objective and make drawings in the circled areas of the Laboratory Report. Look for differences in form, shape, and size.

2. If the Gram stains are indicative of streptococci (Gram-positive cocci in pairs or chains), note their hemolysis pattern on blood agar, enter in the Laboratory Report, and proceed as follows:

   a. With a 1-ml pipet, transfer 0.3 ml of extraction enzyme to each of three small test tubes. Label the tubes: 1 for unknown #1, 2 for unknown #2, and 3 for negative control.

   b. With a sterile loop, remove a single sweep of confluent growth from blood agar culture #1 and transfer the contents to the #1 tube of extraction enzyme. Vortex the tube, then incubate it in 37° C water bath for 30 minutes.

   c. Repeat step b with unknown #2.

   d. Incubate tube #3, the negative control, in the water bath.

   e. Following incubation, add 0.3 ml of sterile physiological saline to each tube. Mix well by vortexing.

   f. Thoroughly mix the two vials of antibody-coated latex bead suspensions (A and B) by manual shaking. Make sure the beads are entirely resuspended.

   g. With a wax marking pencil, inscribe two 14-mm circles on each of four clean glass slides. Label the slides as shown in figure 30.2.

   h. Dispense one drop of Lancefield Group A latex bead suspension near the center of each of the four circles on slides 1 and 3 (see figure 30.2).

Figure 30.2 Procedural arrangement for addition of antigens and antibodies to latex agglutination slides.

A = Lancefield Group A antibody-coated latex bead suspension
B = Lancefield Group B antibody-coated latex bead suspension
1 = Unknown #1 organism-enzyme extract
2 = Unknown #2 organism-enzyme extract
3 = Negative control tube #3
+ = Polyvalent positive antigen control

   i. Repeat step h for Lancefield Group B latex bead suspension on slides 2 and 4.

   j. With a Pasteur pipet, transfer 1 drop of unknown #1 organism-enzyme extract suspension near the center of the left circle on slides 1 and 2. Discard pipet in the hazardous waste container.

   k. Repeat step j with unknown #2 organism-enzyme extract suspension on the right circle of slides 1 and 2.

   l. Add 1 drop of the polyvalent positive control reagent near the center of the left circle on slides 3 and 4.

   m. With a Pasteur pipet, transfer 1 drop of the negative control extraction enzyme (tube 3) near the center of the right circles on slides 3 and 4 (see figure 30.2).

   n. With a stirrer (such as a toothpick), mix the contents of each circle.

   Note: Use a clean stirrer for each circle.

   o. Rock each slide for two minutes either on a mechanical rotator (95–110 rpm) or gently by hand with a rocking motion (see figure 29.1).

   p. Examine each freshly prepared slide for agglutination with a high-intensity incandescent light source.

   Note: Delays in reading agglutination reactions can result in drying of slides, which could render the results uninterpretable.

   q. Report your agglutination test results in table 30.2 of the Laboratory Report.

Note: In reading the slides, you should first compare the agglutination patterns of the positive and negative controls before proceeding to the unknowns.

EXERCISE

30

Laboratory Report: Lancefield Grouping of Pathogenic Streptococci with a Latex Slide Agglutination Test

Results

1. Drawings of unknown Gram-stained bacteria seen with the oil-immersion objective:

Unknown #1

Unknown #2

Blood agar   TS broth

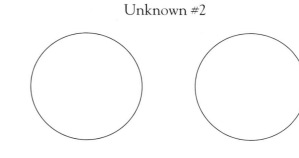

Blood agar   TS broth

2. Describe the type of hemolysis found on blood agar:

Unknown #1:

Unknown #2:

3. Record the latex agglutination reactions (+ or –) in table 30.2.

Table 30.2 Latex Agglutination Reactions with Lancefield Group A and Group B *Streptococcus* Antisera

Test Antigen	Lancefield Group A Antiserum	Lancefield Group B Antiserum
1. Unknown #1		
2. Unknown #2		
3. Negative control		
4. Polyvalent + control		

4. From the above three studies, which unknown did you identify as *S. pyogenes* _____

   and as *S. pneumoniae* _____?

5. Were all of your findings consistent with the literature? If not, describe any inconsistencies observed and if possible provide an explanation.

Questions

1. Why are Group D streptococci not included in the polyvalent positive control?

2. What are some reasons for including positive and negative controls?

3. What advantages are there of the latex slide test over the capillary precipitin test for Lancefield grouping of pathogenic streptococci?

4. *Streptococcus pneumoniae* bacteria often possess cell wall surface antigens that react with Lancefield Group C antiserum (Slifkin and Pouchet-Melvin, 1980). In this event, how would you determine if the positive agglutination result is due to a Group C *Streptococcus* or to *S. pneumoniae*?

NOTES:

EXERCISE

31

Use of an Enzyme-Linked Immunoassay (ELISA) Test for Group A *Streptococcus* Identification

Getting Started

The immunological test used in this exercise is an enzyme-linked immunoassay (or ELISA), which is widely used today for identification of both plant and animal pathogens, including viruses. Plant virus identification examples are cymbidium mosaic virus in orchids and potato virus (PVS) in potatoes. In the clinical setting, it is used for identification of a variety of microbial pathogens since it is a very sensitive test that often requires a simple swab sample from the infected host. The test is routinely used to screen blood for evidence of infection with the hepatitis B virus. It serves as an example of using a combined precipitin-enzyme reaction to achieve an end point.

In this exercise, the ELISA technique is used to identify Group A beta-hemolytic streptococci. There are various modifications of this antigen-antibody immunoassay technique. One of the most sensitive is the **double antibody sandwich method** for cellular antigen detection (figure 31.1*a*), which is used here (see exercise 30 for a related antigen-antibody test). With the double antibody sandwich method, antiserum containing the antibody is first adsorbed on the inner surface of a glass well. Addition of test antigen follows. If the antigen reacts specifically with the adsorbed antibodies it will be retained there when the well is washed free of unbound antigen. Antibody specific for the antigen is then added. If both the antibody adsorbed to the well and the antibody known to be specific for the antigen have reacted with the antigen, the net result is a "sandwich" with the antigen linked to the two antibodies. The reaction becomes visible to the observer only because an enzyme (such as peroxidase or phosphatase) is linked to the second antibody, which elicits a visible color reaction in the presence of a suitable substrate. An indirect ELISA assay, in which the Ag rather than the Ab is adsorbed to the well, is shown in figure 31.1*b*.

In this exercise, you will have an opportunity to evaluate the double antibody sandwich technique for detecting Group A streptococci. The system to be used was developed by Abbott Laboratories, where it is used as a rapid screening and confirmatory test for the detection of Group A streptococci from throat swabs. Since the test does not require living organisms, it may detect some strep on the swab that could fail to grow on culture media plates.

Procedurally, it is first necessary to release cell wall antigens from suspect streptococci using the 5- to 30-minute nitrous acid carbohydrate acid extraction procedure (figure 31.2). The resultant extract is neutralized and transferred to a reaction filter disc coated with a rabbit serum containing strep A antibody (the serum is coated on the filter in such a way that it will not become detached when the filter is washed in a later step to remove unattached antigens and antibodies). During this procedure, strep A antigen if present becomes attached to strep A antibody. Then a solution containing strep A antibody conjugated with the enzyme **alkaline phosphatase** is added to the reaction filter disc. This antibody-enzyme combination conjugates with a free end of the strep A antigen, a part of which, as previously mentioned, is attached to strep A antibody. Next, the coarse prefilter on the reaction disc is removed and discarded. A buffered saline washing solution is then added to the filter disc to remove any unattached antigens and antibodies. Next, a chromogen solution is added, which changes color if alkaline phosphatase is still present. Finally, a solution containing buffered saline is added to stabilize and clarify the color reaction (see figure 31.2 for a schematic diagram).

The color reaction occurs in a vertical and crossing horizontal bar format (figure 31.3). The appearance of both vertical and horizontal bars indicates that the swab, or positive control bacterium, contained Group A strep antigen. A horizontal purple bar alone serves as a procedural control and denotes that Group A strep antigen is not present. A test in which neither bar appears constitutes either a procedural error caused by improper addition of the various reagents or a chemical deterioration of the reagents.

Definitions

Alkaline phosphatase. An enzyme that is active in an alkaline medium. It accelerates the hydrolysis and synthesis of organic esters of phosphoric acid, with the concomitant transfer of phosphate groups to other compounds. In the ELISA test, phosphate released from the added chromogen causes it to change from colorless to colored.

Direct double antibody sandwich. An immunogenic sandwich in which the antigen (or meat) is attached on both ends to two antibodies (the bread).

Objectives

1. To introduce the concept, advantages, and use of an immunological technique (ELISA) employing an antigen-antibody reaction with an enzyme induced colorimetric end point.
2. To learn how to identify Group A hemolytic streptococci with the ELISA technique.

References

Nester et al. *Microbiology: A human perspective*. Chapters 17, 399; 20, 456–457.

Schwabe, L. D., Small, M. T., and Randall, E. L. 1987. "Comparison of TestPack Strep A test kit with culture technique for detection of Group A streptococci." *Journal of Clinical Microbiology* 25:309–311.

Voller, A., Bidwell, D., and Bartlett, A. 1980. "Enzyme-linked immunosorbent assay," 359–371. In N. R. Rose and H. Friedman (eds.), *Manual of clinical immunology*, 2d ed., American Society for Microbiology, Washington, D.C.

Yolken, Robert H. 1985. "Solid-phase enzyme immunoassays for the detection of microbial antigens in body fluids," 949–957. In E. H. Lennette, A. Balows, W. J. Hausler, Jr., and H. Jean Shadomy (eds.), *Manual of clinical microbiology*, 4th ed., American Society for Microbiology, Washington, D.C.

Materials

Positive control: Blood agar streak plate (24 hour, 37° C) of a Lancefield Group A *Streptococcus pyogenes* strain (such as ATCC 19615)

Negative control: Blood agar streak plate (24-hour, 37° C). Some possibilities include an alpha-hemolytic or *viridans* streptococcus (for example, *S. mutans* ATCC 25175)

An Abbott Lab* TestPack Strep A Test Kit (No. 1301-22) that contains 40 tests. For an 80-test kit, order No. 1301-82. A 40-test kit contains:

40 Strep A (rabbit) antibody-coated reaction discs

Reagent 1: 8 ml sodium nitrite (0.5 M)

Reagent 2: 8 ml acetic acid (1.0 M)

Reagent 3: 8 ml TRIS buffer (0.7 M)

Reagent A: 6.5 ml of an anti-Strep A (rabbit) alkaline phosphatase conjugate (0.1 µg/ml)

Reagent B: 45 ml buffered saline (1.0 M)

Reagent C: 6.5 ml chromogen (0.01%)

Reagent D: 45 ml buffered saline (1.0 M)

Also includes extraction tubes and swabs (for obtaining throat swab cultures)

Blood agar plates, 1 per student if you decide to include throat swab cultures

GasPak jar(s) for incubating blood agar plates

Hazardous waste container

A watch or clock with a second hand

*We wish to thank the Abbott Laboratories Diagnostic Division for their help in providing for educational use TestPack Strep A test kits at a reasonable cost.

Figure 31.1 The ELISA technique. (*a*) The double antibody sandwich method and (*b*) the indirect antibody method.

(a) **Double antibody sandwich method**

Antibody is adsorbed to well.

Test antigen is added; if complementary, antigen binds to antibody.

Enzyme-linked antibody specific for test antigen then binds to antigen, forming sandwich.

Enzyme's substrate (■) is added, and reaction produces a visible color change (●).

(b) **Indirect immunosorbent assay**

Antigen is adsorbed to well.

Test antiserum is added; if antibody is complementary, it binds to antigen.

Enzyme-linked anti-gamma globulin (anti-antibody) binds to bound antibody.

Enzyme's substrate (■) is added, and reaction produces a visible color change (●).

Procedure

Safety Precautions: After using the blood agar culture plates, either discard them in the hazardous waste container or autoclave them immediately. Discard any used throat swabs in the hazardous waste container.

Note:

a. All TestPack reagents should be brought to room temperature before beginning the exercise.

b. All bottles should be held vertically so that the reagents are dispensed as free-falling drops.

c. To avoid cross contamination, do not allow bottle tips to contact reaction discs.

d. It is important to use fresh blood agar cultures of both the positive and negative control streptococci.

A pictorial step-by-step scheme of the Procedure below is shown in figure 31.2.

1. Label two extraction tubes as follows: + for positive control and – for negative control,

   Note: You may wish to substitute a throat swab (see ex. 24 for throat swab procedure) for the negative control culture. If so, in as much as some of you may exhibit positive results, the instructor will need to conduct a negative control test with the designated culture, and provide the results to the class.

2. Add 3 drops of Reagent 1 (sodium nitrite) to each extraction tube. The resultant solutions should be pink or purple.

3. Add 3 drops of Reagent 2 to each tube. The resultant solutions should turn yellow signifying an acid pH.

   Warning: Reagent 2 is an irritant, with a pH of 2.2. Avoid contact with the eyes.

4. With a sterile swab, aseptically remove a colony from the positive blood agar plate and place the swab in the + tube, twirl the swab to mix, and let stand for a minimum of 5 minutes but no longer than 30 minutes, with the tube resting in the upright position.

5. Repeat step 4 with a negative control culture.

6. Add 3 drops of Reagent 3 (Tris buffer) to each tube. Twirl swabs to mix. The solutions should turn light pink to purple. A rare specimen may require one more drop of reagent 3 to change color.

7. Express the liquid from each swab by pressing and rotating the fiber portion against the tube wall. The liquid must be thoroughly removed from the swab. Best results are obtained when the liquid reaches or exceeds the line on the tube. Discard the swab in the hazardous waste container.

   Note: Unless you have had previous experience, steps 8 thru 12 should be run with one sample at a time. If both samples are run concurrently, it is important to complete each step for both reaction discs before proceeding to the next step. The liquid obtained in step 7 above is stable for 72 hours when refrigerated in a covered tube.

8. Pour the entire contents of the + tube into the center of the reaction filter (**Note:** Do not remove the purple prefilter.) and allow it to soak through the filter (approximately 10 seconds).

9. Add 3 drops of Reagent A (Anti-strep A alkaline phosphatase) to the filter and incubate 1 minute. **Note:** Complete the next step within one minute of the Reagent A addition.

10. Remove and discard the purple prefilter in the hazardous waste container. Wash the contents of the filter by filling the dispenser dropper of Reagent B (buffered saline) to the white line, and adding it to the filter surface of the reaction disc. Allow complete flow through.

11. Add 3 drops of Reagent C (the chromogen) to the filter surface of the reaction disc.

    Note: Allow color to develop for 2 minutes, which is the time necessary for an accurate result.

12. Wash the filter by filling the dispenser dropper to the white line with Reagent D (buffered saline), and adding it to the filter surface of the reaction disc. Allow solution to flow through completely.

Figure 31.2 Step-by-step scheme of ELISA test for Group A *Streptococcus* identification.

Figure 31.3 Interpretation of TestPack Strep A ELISA results. Reproduction of *Interpretation of Abbott TestPack Strep A ELISA Results* has been granted with approval of Abbott Laboratories, all rights reserved by Abbott Laboratories.

A Positive Sign (+) indicates that the specimen contains group A strep. Light gray to dark purple, darker than the background, on the vertical line should be interpreted as a positive result even if it has less color than the horizontal line.

A Negative Sign (–) indicates that a properly extracted specimen should be considered negative for group A strep. If the test result is inconsistent with clinical presentation, it is possible that an improper sample was collected. A follow-up throat swab should be obtained.

Tests that do not result in a **Positive (+)** or a **Negative (–)** sign may indicate improper addition of reagents or deterioration of reagents. The patient should be retested.

13. Repeat steps 8 through 12 for the negative control tube.
14. Read your TestPack Strep A results using figure 31.3 for interpretation. Enter your results in table 31.1 of the Laboratory Report.

15. Observe the type of hemolysis reaction on your blood agar plates and enter your results in table 31.1 of the Laboratory Report.

Note: You may wish to consult exercise 24 for an interpretation of the hemolysis reactions.

EXERCISE

31

Laboratory Report: Use of an Enzyme-Linked Immunoassay (ELISA) Test for Group A *Streptococcus* Identification

Results

1. Following the interpretive guides (see Procedure, steps 14 and 15) fill in the necessary blanks in table 31.1.

Table 31.1   Group A *Streptococcus* Identification: ELISA and Hemolysis Reactions on Blood Agar

Sample Description	Hemolysis Reaction	ELISA Test Result
Control culture (+)		
Control culture (−)		
Throat swab		

2. Discuss your results and their significance.

3. What future studies might be suggested from your results?

Questions

1. What is the function of Reagent 1 (sodium nitrite) in the ELISA test?

2. What are three factors that can limit the value of the ELISA test?

3. In the Abbott Strep A diagnostic technique, the horizontal control bar elicits a purple color regardless of whether or not a Group A *Streptococcus* is present. What immunological explanation might there be for such a result?

4. Do you know any examples of other ELISA tests used for identification of pathogenic microbes?

5. Compare the ELISA test with the latex agglutination test for identification of Group A streptococci with respect to the following:

a. Spectrum of organisms

b. Extraction of cell wall antigens

c. Nature of the antigen-antibody reaction

NOTES:

EXERCISE

An Ouchterlony Double Immunodiffusion Test for *Candida* Identification

Getting Started

As stated in the Part Four Introduction, for non-particulate antigens a **precipitin reaction** will occur with the appropriate antibody, and can sometimes be observed visually. For example, in the Ouchterlony test, a precipitin line is formed in an agar base and can often be seen visually without the aid of additional light or staining. The test is widely used for **serodiagnosis** of fungal diseases (Kaufman and Reiss, 1985), although use of ELISA (see exercise 31) and radioimmunoassays are the subject of serious investigation for diagnosis of candidiasis and aspergillosis (de Repentigny and Reiss, 1984).

With the Ouchterlony procedure, soluble antigen and serum solutions (antibodies) are placed in separate wells of an agar base. The reactants diffuse from the wells and form thin, white precipitin line(s) where they meet in optimum proportions (figure 32.1). Since both immunoreactants move in this system, it is known as a **double immunodiffusion test.** In a related system, the Oudin technique, diffusion occurs only in a single dimension (figure 32.1*b*).

According to the **lattice theory** depicted pictorially in figure 32.2, the precipitin lines form best where an excess of antibody relative to antigen is present. When the Ab/Ag ratio is less than 1, soluble complexes rather than precipitates occur (figure 32.2 *d–f*). In some instances, more than one precipitin line will form in the agar, because the antigen preparation contains more than one antigen; thus, each band formed represents one antigen-antibody reaction.

The curvature of the precipitin line can provide information about the molecular weight of the antigen, providing the antigen and antibody are present in nearly equal amounts. The precipitin line appears straight if the antigen and antibody have about the same molecular weight. If the antigen has a higher molecular weight, the line is

Figure 32.1 Some commonly used systems for gel diffusion precipitin reactions. (*a*) Double diffusion in two dimensions (Ouchterlony technique) in which diffusible antigen and antibody solution are placed in separate wells cut in an agar plate. Direction of diffusion is shown by *arrows*. (*b*) Single diffusion in one dimension (Oudin technique) using a soluble diffusion antigen layered over an antibody contained in an agar gel column. Black lines are opaqued precipitin bands. Reproduced with the permission of Meridian Diagnostics, Inc., Cincinnati, Ohio 45244.

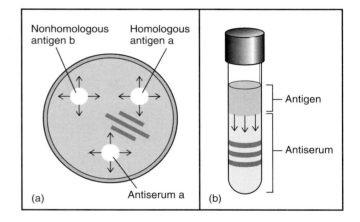

concave toward the antibody well; if the antigen is of lower molecular weight, the line is concave toward the antigen well. These relationships are derived from the principle that the rates of diffusion of any mixture increase with concentration and decrease with molecular weight.

Finally, antigens or antibodies can be tested for **identity** by placing a test well of the substance in question adjacent to the wells of a known system. If the antigen-antibody complexes are identical, the precipitin lines form an unbroken line of identity with the known system (figure 32.3*a*). **Partial identity** and **nonidentity** reactions are also possible (figure 32.3*b* and *c*, respectively). A partial identity reaction occurs when certain components of the antigens (or antibodies) are identical and others are not. The **spur** represents the components that are unrelated. A nonidentity

Figure 32.2 (*a–f*) Use of the lattice theory to postulate the proper ratios of Ab/Ag necessary to form a precipitin line.

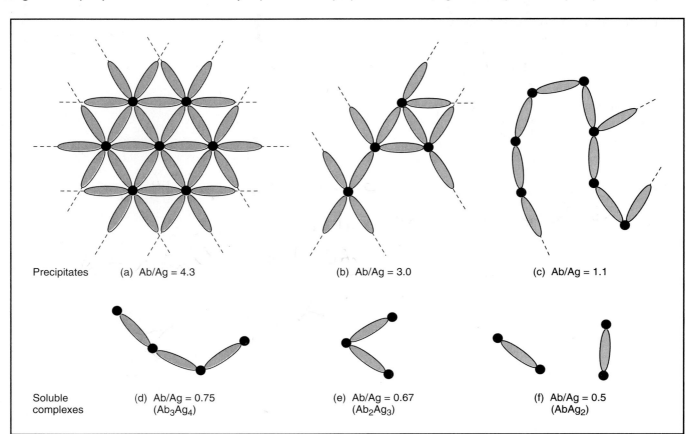

Precipitates   (a) Ab/Ag = 4.3   (b) Ab/Ag = 3.0   (c) Ab/Ag = 1.1

Soluble complexes   (d) Ab/Ag = 0.75 (Ab_3Ag_4) (e) Ab/Ag = 0.67 (Ab_2Ag_3) (f) Ab/Ag = 0.5 $(AbAg_2)$

reaction will occur when the antigen-antibody complexes are different. The resulting "X," or cross reaction, indicates that two unrelated complexes are present.

The test organism chosen to illustrate the principles of Ouchterlony double immunodiffusion is a sometimes pathogenic yeast, *Candida albicans*. It occurs as a normal resident of body flora but at times can cause **opportunistic infections** in the vagina (where it causes vaginitis) and in the oral cavity (where it causes thrush, especially in **immunosuppressed patients).** In this study, we use a prepared *Candida* extract as an antigen source, rather than living cells.

Definitions

Double immunodiffusion test. An immunological test in which both the antigen and the antibody move in the reaction system.

Immune response. Specific response to a foreign antigen characterized by the production of humoral antibodies or immune cells.

Immunosuppressed patient. A patient whose normal immune response is suppressed due, for example, to the presence of alcohol in the body or a disease, such as cancer.

Figure 32.3 Double diffusion precipitin reactions observed in agar gel plates:

1. Antigen solutions with one or more antigenic components. Antigen solution A contains two distinct antigen components as indicated by precipitin lines a_1 and a_2.
2. Lines of identity (fusion). (*a*) Antigen A solution contains a component antigenically identical with antigen solution B. For example, A shows a line of identity by fusion and thickening of precipitin lines a_2 with b_2.
3. Lines of partial identity (spur). (*b*) Antigen C contains a component (c_2) that shares at least one antigenic determinant with antigen A, but A contains at least one antigenic determinant that is nonidentical to C. The evidence for the partial identity of antigens A and C is the merging and thickening of the C precipitin line with the A precipitin line, but the A precipitin line has a spur that overlaps the C precipitin line.
4. Lines of nonidentity (cross). (*c*) In this instance, the precipitin lines do not fuse but cross because the antigens A and D are not related chemically to one another.

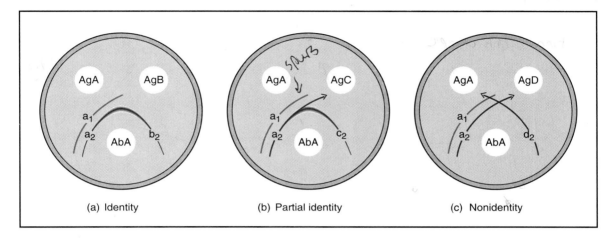

(a) Identity   (b) Partial identity   (c) Nonidentity

Lattice theory. A theory based on a framework formed by crossing soluble antigens and antibodies in a diagonal manner such that when present in the correct ratio of Ab to Ag, an aggregate forms, which when sufficiently large, precipitates out of solution in the form of a visible complex.

Opportunistic infection. An infection in which organisms normally nonpathogenic or weakly pathogenic, are able to cause disease in an immunosuppressed patient.

Precipitin reaction. A reaction in which an antibody is able to combine with an identifiable soluble antigen and cause aggregation and eventual precipitation out of solution.

Serodiagnosis. A diagnostic test or tests employing antigen-antibody reactions.

Objectives

1. To familiarize you with some of the principles of Ouchterlony double immunodiffusion, and to provide an opportunity for you to conduct and observe some of these principles at work in the laboratory.
2. To perhaps suggest some improvements in methodology once you have conducted this exercise.

References

Kaufman, L., and Reiss, E. 1985. "Serodiagnosis of fungal diseases," 924–944. In E. H. Lennette, A. Balows, W. J. Hausler, Jr., and H. Jean Shadomy, eds., *Manual of clinical microbiology*, 4th ed. American Society for Microbiology, Washington, D.C.

Nester et al. *Microbiology: A human perspective*. Chapters 17, 395–396; 19, 445–446; 22, 510, 517; 24, 563–564.

de Repentigny, L., and Reiss, E. 1984. "Current trends in immunodiagnosis of candidiasis and aspergillosis." *Review Infectious Diseases*, 6:301–312.

Materials

A Meridian Diagnostics Inc. Test Kit (120100),* which contains:

0.5 ml *Candida* ID antigen

1.0 ml of diluted anti-*Candida* ID control serum (for preparation see Appendix 5, Notes to Instructor)

Immunodiffusion agar plates with recording forms, 12

Capillary pipets with bulb, 80

Additional bulbs for use with capillary pipets

Moist chamber: a Petri dish with a tight fitting cover containing moist filter paper or paper toweling is satisfactory provided the ID plates remain level and hydrated during the incubation period.

A reading light, such as a bright light source wherein the ID plate is held in the light beam at a 45° angle against a dark background. Commercial light boxes providing indirect light against a dark background are also available.

Reagent quality water

*We wish to thank Meridian Diagnostics, Inc., for their willingness to provide outdated usable Immunodiffusion kits at a reasonable cost (see Notes to Instructor).

Procedure

Note: Refer to figure 32.4 of the Laboratory Report for a description and numerical designation of the test well pattern of the ID agar plate.

1. Using a capillary pipet with an attached bulb, fill the pipet approximately ¾ full with *Candida* antigen.
2. Fill the center well (#7 on figure 32.4) with *Candida* antigen.

Note: Take care to avoid overflow of antigen solution from the well.

3. Using a fresh capillary pipet, repeat steps 1 and 2 with positive control antiserum. Fill well #1 with positive control serum. If you wish to observe a line of identity, also fill well 2 or 6 with positive control serum (see color plate 24).
4. If negative control sera or positive sera from patients known to have a *Candida* infection are available, they can be added to other external wells of the ID agar plate. For example, if positive serum is available from a patient it should be placed in a well adjacent to one containing positive control antiserum. Such an arrangement will enable you to determine lines of identity, partial identity, or nonidentity (see color plate 24).
5. Identify the contents of each well on the report sheet. Also make a note in which wells, if any, the antiserum was diluted, and the amount of the dilution.
6. Place your name or initials and date on the ID plate cover. Incubate it in a moist chamber at room temperature for 24–48 hours.
7. After 24 to 48 hours incubation, read and record the ID bands in table 32.1 of the Laboratory Report. A light source is preferred for observing the nature of the bands. Particular attention should be paid to the orientation of the bands in relation to control serum bands. A smooth junction of the bands is indicative of an identity reaction (see color plate 24). If antiserum from a patient were included in this test, you would also look to see if the control band was bent toward a position in front of the patient well. This would indicate patient antibody at a low titer.
8. Interpretation of the Test
   a. A band of identity with a known positive control indicates patient antibody against the antigen in question. In general, an identity reaction against a given antigen is indicative of active or recent past infection.
   b. Partial identity reactions are regarded as positive for antibody against the antigen only if no other identity reaction is

present on the plate. Partial identity reactions are also indicators of possible disease.

c. Nonidentity reactions are regarded as a negative test, and may also be apparent when the disease state is caused by a mycotic agent other than the one tested.

d. The greater the number of precipitin lines observed, the greater the likelihood of severe candidiasis including disseminated, near terminal candidiasis and endocarditis. In arriving at a diagnosis, the test results should be interpreted together with all other lab and clinical data, including treatment history.

9. Limitations of the Test. The high rate of negative serologic tests observed among culturally demonstrable cases limits the predictive value of a negative test. Positive test results must be confirmed culturally.

NOTES:

antigen - markers on the outside of cell (♂) proteins (glycoproteins)

B-lymphocytes = produce immunoglobulins (antibodies) ÷ Y-shaped

each antibody is antigen-specific (can only recognize certain antigens

when antibody binds to B-lymphocytes, more free antibodies are released.

Agglutination : Ab complexed - particulate $\overset{Ag}{}$

Precipitation : complex between Ab and nonparticulate (soluble) Ag
important ratio`` Ab 4 : 3 Ag
lattice

Immunodiffusion - Ouchterlony wells

center hole { antibodies

antihorse

antibovine

antigens
horse
bovine
swine
horse/bovine
saline

EXERCISE

32

Laboratory Report: An Ouchterlony Double Immunodiffusion Test for *Candida* Identification

Figure 32.4 Test well pattern of the immunodiffusion agar plate.

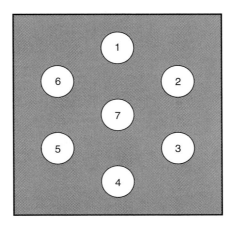

Results

Date ID plate incubated _____ Date observed _____

1. Make drawings on figure 32.4 of all observed precipitin lines.
2. Fill in the necessary blanks in table 32.1.

   Note: Indicate under Well Description in which wells, if any, the serum was diluted.

Table 32.1 Ouchterlony Fungal Immunodiffusion Analysis Form

Well No.	Well Description	Reading Observed
1	Positive control serum	
2		
3		
4		
5		
6		
7	Antigen:	

Note: Test is invalid if positive control is negative after 24-hour incubation.

3. Discuss your results and their significance.

4. What future studies might be suggested from your results?

Questions

1. Interpret the following *Candida* immunodiffusion bands with respect to the possibility of having an active *Candida* infection:

(a) — Positive serum control
— Patient's serum
— *Candida* antigen

(b)

(c)

(d)

2. Interpret the relative antigen (Ag) and antibody (Ab) molecular weights from the following immunodiffusion bands:

a. ID test using equal concentrations of Ag and Ab.

o – – Ab
_____
o – – Ag

b. ID test using a greater concentration of Ag than Ab.

o – – Ab
_____
o – – Ag

c. ID test using equal concentrations of Ag and Ab.

o – – Ag

o – – Ab

3. Circle the Ag (•) Ab (○) diagram in which the best precipitin line is possible:

4. What advantages does Ouchterlony double immunodiffusion precipitin analysis have over Oudin single immunodiffusion analysis?

Public Health and the Role of

Microbiology in the Prevention and

Control of Communicable Diseases

/*ntroduction* to the Prevention and Control of Communicable Diseases

Figure I.10.1 (*a*) The John Snow pub in London where epidemiologists go to celebrate the heroics of John Snow's early epidemiological efforts to help stem a cholera epidemic. (*b*) A replica of the pump, with pump handle attached, a monument dedicated to Dr. Snow in July 1992. At the time of the epidemic, he was so convinced that the disease was being carried by water from the pump that he had the pump handle removed. Koch isolated and identified the cholera vibrio about 30 years later. Courtesy of Kathryn Foxhall

(a)

(b)

Communicable, or infectious, diseases are transmitted from one person to another. Transmission is either by direct contact with a previously infected person, for example by sneezing, or by indirect contact with a previously infected person who has contaminated the surrounding environment.

A classic example of indirect contact transmission is an epidemic of cholera that occurred in 1854 in London. During a 10-day period, more than 500 people became ill with cholera and subsequently died. As the epidemic continued, John Snow and John York began a study of the area and were able to prove by **epidemiological methods** only (the bacteriological nature of illness was not known at that time) that the outbreak stemmed from a community well on Broad Street known as the Broad Street Pump* (figure I.10.1).

Then they discovered that sewage from the cesspool of a nearby home was the pollution source, and that an undiagnosed intestinal disorder

*Snow, John: "The Broad Street Pump," in Roueche, Berton (ed.): *Curiosities of medicine*. Berkely, Medallion, ed., New York, 1964.

had occurred in the home shortly before the cholera outbreak. They also learned that neighboring people who abstained from drinking pump water remained well, whereas many of those who drank pump water succumbed to cholera.

Today the incidence of cholera, typhoid fever, and other infectious diseases rarely reach **epidemic** proportions in those countries that have developed standards and regulations for control of environmental reservoirs of infection. The major reservoirs are water, food, and sewage.

The importance of epidemiology in tracing the source of an infectious disease is demonstrated in laboratory exercise 33. The exercise involves the use of a method for detecting a specific *Staphylococcus* strain on various body parts of the student class. If all members of a class carry the same strain, an epidemic is likely to exist.

The subject of public health sanitation is presented in exercise 34 (water microbiology).

All of the above studies come under the surveillance of public health agencies responsible for prevention and control of communicable diseases. Among these are the Department of Health, Education, and Welfare at the federal level, which has cabinet status, conducts preventive medicine research, provides hospital facilities for service men and women, and gives financial assistance to state and local health departments. Additionally, the Centers for Disease Control (CDC) in Atlanta play an important role in the prevention and control of communicable disease. Also, all states and other local government agencies perform important public health services.

Perhaps the most important international health organization is the World Health Organization (WHO) headquartered in Geneva, Switzerland. WHO distributes technical information, standardizes drugs, and develops international regulations important for the control and eradication of epidemic diseases. For example, smallpox, which was once a widespread disease, is virtually nonexistent today.

Finally, there are voluntary health organizations that help in some of the above causes mentioned previously. Among these are privately endowed foundations (such as the Commonwealth, Ford, and Rockefeller Foundations), the American Red Cross, the American Public Health Association, and religious orders of many denominations.

Definitions

Epidemic. The occurrence in a community or region of a group of illnesses of similar nature, clearly in excess of normal expectancy.

Epidemiological methods. Methods concerned with the extent and types of illnesses and injuries in groups of people, and with the factors that influence their distribution. This implies that disease is not randomly distributed throughout a population, but rather that subgroups differ in the frequency of different diseases.

EXERCISE

33

Epidemiology: A *Staphylococcus* Carrier Study

Getting Started

Nosocomial (hospital-acquired) infections and epidemics are common—they account for a considerable proportion of infections in hospital patients. This is not surprising, since these patients comprise a highly susceptible group of people. Moreover, the hospital is an environment in which procedures or treatment may have the effect of reducing normal body resistance to infection, and in which the use of antibiotics has fostered the development of drug resistant strains. What may come as a surprise to health science students is the degree to which hospital personnel themselves can be carriers and transmitters of pathogens that may also be resistant to antibiotics in many cases. In this exercise, you find out how many of the students in your class are carriers of the potential pathogen *Staphylococcus aureus*, and on what part of the body these bacteria occur.

Since penicillin and other antibiotics have been used increasingly over the last 50 years to treat infections, many hospital-isolated strains are resistant to one or more of the antibiotics to which the original *Staphylococcus* strains were formerly susceptible. Comparison of your antibiogram (antibiotic susceptibility pattern) to that of the reference strain demonstrates how these antibiograms have changed with the use of antibiotics.

When a number of patients in a given hospital have *Staphylococcus aureus* infections, it is often difficult to determine if a true epidemic exists, because so many people are routine carriers of these organisms on their body. However, by identifying different strains within the single species, it is possible to determine the existence of a true epidemic arising from a single source. If all isolated organisms come from the same strain, then an epidemic undoubtedly exists; if they are of different strains, then the infections are probably not directly related.

Variations in susceptibility to antibiotics (the antibiogram), and in the production of **hemolysin** and **coagulase** can be studied to demonstrate different strains of *S. aureus*. In clinical studies, susceptibility of staphylococci to infection with different bacteriophages can be used for strain differentiation. The greater the number of characteristics studied, the more accurate strain identification becomes.

In this exercise, you prepare streak plates on mannitol-salt agar (a **selective medium** used for isolation of pathogenic staphylococci) of swabs taken from three areas of the body: the throat, nose, and skin. If a mannitol-fermenting colony is isolated from any source, you test an isolate for its ability to produce hemolysin and coagulase, and for resistance to six antibiotics to which the typical reference strain of *S. aureus* is susceptible.

The results of the above studies are then reported in table 33.1 of the Laboratory Report. From this table an Information Sheet is prepared and turned in to the instructor for tabulation with those of the other students. A summary of the tabulated data appears in table 33.2. With the latter data, you will be able to answer the epidemiological questions suggested in the Question section of the Laboratory Report.

Definitions

Coagulase. An enzyme secreted by *S. aureus* that clots plasma. It contributes to virulence and to forming a fibrin wall that surrounds staph lesions.

Hemolysin. Any biological agent capable of lysing red blood cells with concomitant release of hemoglobin. Examples include certain exotoxins and complement related reactions.

Selective medium. A growth medium designed to favor the growth of certain microbes and inhibit the growth of undesirable competitors.

Objectives

1. To enable you to participate actively in an epidemiological investigation, wherein you can use your microbiological skills to obtain new information (see Laboratory Report Questions 1 through 4).
2. To test three hypotheses about staphylococcal carriers (see Laboratory Report Questions 5 through 7).

References

Mausner, J. S, Kramer, Shira et al. *Epidemiology, an introductory text*. 2nd ed. Philadelphia: W. B. Saunders Co., 1985.

Nester et al. *Microbiology: A human perspective*. Chapter 18.

Materials

Per team of two students

TS broth 24-hour 37° C reference culture of *S. aureus* with its antibiogram checked previously by Instructor.

Mannitol salt agar plates with phenol red indicator, 7

Sterile swabs, each in a sterile test tube, 9

Tongue depressors, 2

Tubes of sterile water, 2

Blood agar plates, 2

Tubes of TS broth, 4

Tubes containing 0.5 ml of coagulase plasma, 2

Sterile Pasteur pipets, 3

Mueller-Hinton agar, 3 plates

A dropper bottle containing *fresh* 3% hydrogen peroxide

Forceps, 2

Antibiotic discs, 3 of each of the following:
penicillin, 10 µg erythromycin, 15 µg
streptomycin, 10 µg tetracycline, 30 µg
sulfanilamide, 300 µg chloramphenicol, 30 µg

Procedure

Safety Precautions: *S. aureus* can cause wound infections.

First Session

1. Each student assembles three mannitol agar plates; label one plate "nose," another "throat," and the last plate "skin." Write your name on each plate.
2. Each student labels three swabs containing test tubes "nose," "throat," and "skin," respectively.
3. Using aseptic technique, have your partner moisten the swab from your tube labeled "nose" in a tube of sterile water, sponge it almost dry against the inside of the tube, and insert it into your anterior nostril by gentle rotation, and return the swab to the empty tube.
4. Repeat step 3 with the swab labeled "throat"; with a tongue depressor, depress the tongue, then rub the swab firmly over the back of the throat and tonsillar region (see figure 24.2). Return the swab to the empty tube.
5. Repeat step 3 with the swab labeled "skin"; rub it over the skin surface located between your fingers and finger tips; return the swab to the empty tube.
6. Streak each swab over ⅓ of the surface of the appropriately labeled mannitol salt agar plate. Discard used swabs in a hazardous waste container. Continue to streak with a loop to obtain isolated colonies.
7. Repeat steps 3 through 6 for your partner.
8. Streak a loopful of the *S. aureus* reference culture on the surface of a mannitol salts agar plate (only one reference culture per team is necessary).
9. Invert plates and incubate at 37° C until next session.

Second Session

1. Examine the reference strain mannitol salt agar plate and note the appearance of typical *S. aureus* colonies (usually large, opaque colonies); any with mannitol fermentation show yellow halos (see color plate 25).

Look for similar colonies on your nose, throat, and skin mannitol salt agar plates. Record their mannitol fermentation results (+ or –) in part 1 of the Laboratory Report.

2. Select several typical appearing staph colonies from one or more of your body isolate plates and test for catalase production (see figure 24.3 for test method). Include a positive control from your reference mannitol salt agar plate.

Note: A positive catalase test is only suggestive of the presence of S. *aureus*, since all staphylococci exhibit a positive catalase test. If all the colonies from your body culture plates are catalase negative, state so in the Laboratory Report, and fill out and turn in the Information Sheet to your instructor. This occurrence completes your laboratory portion of the study.

3. Twenty-four hours before the next laboratory session, subculture a colony from your body isolate mannitol salt agar plate that is catalase positive. Also subculture the S. *aureus* reference strain. They should be subcultured in two tubes of TS broth and on a plate of blood agar. Both your body culture and the reference strain can be streaked on one blood agar plate by dividing the plate into halves. Label the tubes and plates, and incubate at 37° C for 24 hours.

4. Determine the coagulase activity of another catalase positive body plate colony, and the reference colony culture. Use the method outlined in exercise 23. Record your results in table 33.1.

Note: If you have only one catalase positive colony on your chosen body plate, delay the coagulase test until the next session and then use a colony from the blood agar plate for the coagulase test.

Third Session

1. Record the presence or absence of hemolysis on the blood agar plates in table 33.1. Compare it for similarity with the reference strain of S. *aureus*.

2. If the coagulase test is negative, report your culture as negative for S. *aureus*. Fill out

the Information Sheet and turn it in to your instructor.

3. If the coagulase test is positive, set up an antibiotic susceptibility test. The six antibiotic discs to be used are listed in the Materials section. The inocula to be used are the three TS broth cultures prepared in the second session. The procedure is as follows:

a. With a glass-marking pencil, divide the bottom of a Petri dish containing Mueller-Hinton agar into six pie-shaped sections (see figure 14.3b). Repeat the marking procedure with the remaining two dishes of Mueller-Hinton agar. Label one dish Reference culture. The remaining two dishes represent you and your partner's body cultures.

b. Inoculate the reference plate by moistening a sterile swab with the TS broth reference culture, and spreading it uniformly over the plate surface by moving the swab back and forth in all directions (see figure 14.3a). Repeat the inoculation procedure with the remaining two plates using your two TS broth cultures as inocula.

Note: In order to have some semblance of validity, the procedures used to prepare and observe the antibiotic susceptibility test should be similar to that used in exercise 14. Use table 14.2 to interpret your results.

c. Assemble the discs, and record the code of each in part 3 of the Laboratory Report to ensure correct interpretation of test results.

d. Heat-sterilize forceps by dipping in 95% alcohol and flaming. Air cool.

e. Remove one disc aseptically from container. Place gently in the center of one pie-shaped section of the reference plate culture (see figure 14.3b). Tap disc gently with forceps to fix it in position on the agar surface.

f. Continue placing the remaining five discs in the same way. Make certain that you sterilize the forceps after placing each disc, since there is a possibility of contaminating stock vials with resistant organisms or even occasionally with drug-dependent bacteria!

g. Repeat the above procedure with your two body culture plate.

h. Invert and incubate the plates at 35° C for 48 hours.

Fourth Session

1. Using a mm ruler, measure the diameter of the zone of inhibition around each antibiotic disc, and record their diameters in table 33.1. Next consult table 14.2 and determine from the latter information if the cultures are susceptible (S) or resistant (R) to the antibiotic in question. Record S or R in the appropriate square of table 33.1.

Note: The reference culture is expected to be susceptible to all six antibiotics.

2. Fill out the Information Sheet of the Laboratory Report. Tear out and return it to the instructor for tabulation of data to be inserted in table 33.2.

EXERCISE

Laboratory Report: Epidemiology:
A *Staphylococcus* Carrier Study

Results

1. Record the initial results you obtained from your three body cultures on mannitol-salt agar.

Note: At least one colony should show good colony growth with a yellow color change indicating a positive test for mannitol fermentation.

Record each culture as mannitol positive (+) or negative (–):

Nose _____   Throat _____   Skin _____

2. Catalase test results (+ or –):

Nose _____   Throat _____   Skin _____

3. Record the code of each antibiotic disc:

Penicillin _____   Erythromycin _____

Streptomycin _____   Tetracycline _____

Sulfanilamide _____   Chloramphenicol _____

4. Record results obtained with the one strain you chose to study in table 33.1. Also indicate deviation of any of the results obtained for your strain from the reference strain.

Table 33.1   Test Results Obtained with Reference *S. aureus* Culture and Mannitol-Salt Agar Positive Body Culture

| Isolate Source Tested | Mannitol Fermentation (+ or –) | Beta-hemolysis (+ or –) | Coagulase (+ or –) | Antibiotic Susceptibility (S or R) | | | | | |
				Peni-cillin	Strepto-mycin	Tetra-cycline	Chloram-phenicol	Erythro-mycin	Sulfanil-amide
Your Strain									
Reference Strain									
Deviation from Reference Strain									

NOTES:

Name _____ Date _____ Section _____

Instructor Information Sheet for Tabulation of *Staphylococcus* Carrier Study

Which sources showed growth and fermentation on mannitol-salt agar?

Throat ____   Nose ____   Skin ____

Do you work in a clinical setting this quarter?

Yes ____   No ____

If so, where? _____

Have you taken antibiotics this quarter?

Yes ____   No ____

If so, which? _____

Isolate Source Tested	Mannitol Fermentation (+ or −)	Beta-hemolysis (+ or −)	Coagulase (+ or −)	Antibiotic Susceptibility (S or R)					
				Peni-cillin	Strepto-mycin	Tetra-cycline	Chloram-phenicol	Erythro-mycin	Sulfanil-amide
Your Strain									
Reference Strain									
Deviation from Reference Strain									

NOTES:

5. Based on the information for the class given to you by your instructor, complete table 33.2.

Table 33.2   Classroom Summary of *S. aureus* Epidemiological Study

	Health Professionals (students working in clinical setting)		General Population (students not working in clinical setting)		Total Class Results	
	Number	% of sample	Number	% of sample	Number	% of sample
S. aureus carriers (positive cultures)						
Noncarriers (negative cultures)						
Total						
Carriers of susceptible strains						
Carriers of resistant strains						
Total						
Carriers of strainlike reference strain						
Carriers of deviant strains						
Total						
Different *types* of deviant strains						

Total number and % of mannitol positive strains from:

Throat _____ Nose _____ Skin _____ Total _____

(**Note:** This number might be larger than the class, because any student might have any number of sources positive.)

Questions

1. What proportion of students are carriers of the potential pathogen *Staphylococcus aureus*?

2. What parts of the body harbor this pathogen, and which parts carry it most commonly?

3. To what extent has *S. aureus* acquired resistance to antibiotics to which it was originally susceptible?

4. How many different strains can be isolated from the student population that deviate from the "typical" *S. aureus?*

5. Do working health professionals have a higher carrier rate than the general population?

6. Do the strains of S. *aureus* carried by health professionals have a higher proportion of resistant strains than those isolated from the general population?

7. Is there a greater number of different strains of S. *aureus* among health professional carriers than among the general population?

NOTES:

EXERCISE

34

Bacteriological Examination of Water: Multiple-Tube Fermentation and Membrane Filter Techniques

Getting Started

Water, water, everywhere Nor any drop to drink

The Ancient Mariner
Coleridge, 1796

This rhyme refers to seawater, undrinkable because of its high salt content. Today the same can be said of freshwater supplies, polluted primarily by humans and their activities. A typhoid epidemic, dead fish on the beach, and the occurrence of a red tide are all visible evidence of pollution. Primary causes of pollution include dumping of untreated (raw) sewage and inorganic and organic industrial wastes, and fecal pollution by humans and animals of both fresh and groundwater. In the United States, sewage and chemical wastes are in regression largely as a result of the passage of federal and local legislation requiring a minimum of secondary treatment for sewage and the infliction of severe penalties for careless dumping of chemical wastes.

Fecal pollution by humans and animals is more difficult to control particularly as the supply of water throughout the world becomes more critical. In some parts of the world, particularly in Third World countries, it is estimated that over 12,000 children die every day from diseases caused by waterborne fecal pollutants. Examples of such diseases are cholera, typhoid fever, bacterial and amoebic dysentery, and viral diseases such as polio and infectious hepatitis. Most of the inhabitants of these countries are in intimate contact with polluted water because they not only drink it, but also bathe, swim, and wash their clothes in it.

The increased organic matter in such water also serves as a substrate for anaerobic bacteria, thereby increasing their numbers in relation to the **aerobic** bacteria originally present (figure 34.1). Nuisance bacteria, such as *Sphaerotilus natans*, a large, rod-shaped organism that grows in chains and forms an external sheath

Figure 34.1 Diversity of microbes found in pure to very polluted water. Note the change from aerobic to anaerobic microbes as the water becomes more polluted. Courtesy of Settlemire and Hughes. *Microbiology for Health Students,* Reston Publishing Co., Reston, Virginia.

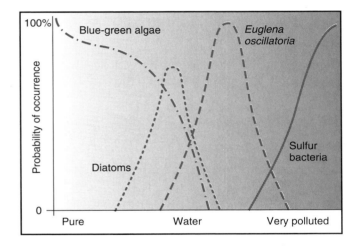

(figure 34.2) are able to adhere to the walls of water pipes. This can eventually cause fouling of the pipes, thereby reducing the carrying capacity of the pipes.

Fortunately, microbes are also beneficial in water purification. In smaller sewage treatment plants, raw sewage is passed through a slow sand filter, wherein microorganisms present in the sand are able to degrade (metabolize) organic waste compounds before the effluent is discharged. Sewage effluent is finally chlorinated to futher reduce fecal microbial contaminants. The development of sewage treatment plants and the control of raw sewage discharge reduced the annual typhoid fever death rate in the United States from about 70 deaths per 100,000 population to nearly zero. However, the potential danger of pollution is always present. In 1973, an epidemic outbreak of typhoid fever occurred in Florida, and in 1975 residents of the city of Camas, Washington were inundated by intestinal disorders that were traced to fecal pollution of the water supply by beavers infected with the protozoan *Giardia lamblia* (figure 34.3).

Figure 34.2 *Sphaerotilus* species, a sheathed bacterium that often produces masses of brownish scum beneath the surface of polluted streams. Phase-contrast photomicrograph courtesy of J. T. Staley and J. P. Dalmasso

Bacterial
cells

Granules of
polyhydroxybutyrate

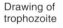
10 μm   Sheath   Septa

Figure 34.3 Human fecal specimen illustrations showing trophozoite (*a*) and cyst (*b*) forms of *Giardia lamblia,* a waterborne protozoan pathogen that inhabits the intestinal tract of certain warm-blooded animals. The cyst form is resistant to adverse environments and is the form released with fecal material. (*a*) Pear-shaped trophozoite. Electron microscopy, iodine-stained, magnification × 2,900, 1 micron = 2.9 mm. Note flagellate appendages for locomotion. (*b*) Smooth ovoid cysts. Scanning electron microscopy, magnification × 2,900, 1 micron = 2.9 mm. Cysts are embedded in a mat of debris, bacteria, and fecal material. (*a*) © J. Paulin/Visuals Unlimited (*b*) From D. W. Luchtel, W. P. Lawrence, and F. B. De Walle, "Electron Microscopy of Giardia lamblia Cysts," *Applied and environmental microbiology,* 40:821–832, 1980. © American Society for Microbiology.

Drawing of
trophozoite

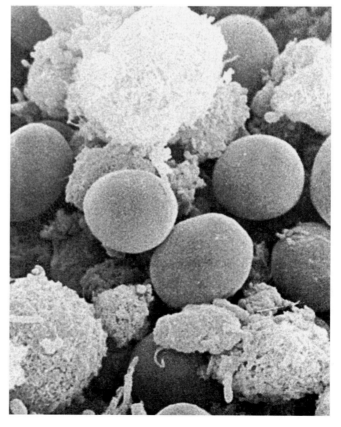

(a)

(b)

Two microbiological methods commonly used for determining whether a given sample of water is polluted are:

1. The determination of the total number of microorganisms present in the water. This **plate count method** provides an indication of the amount of organic matter present. In view of the great diversity in microbial physiology, no single growth medium and no single set of cultural conditions can be found that satisfy universal microbial growth. Hence, a choice had to be made. Experience taught that standard plate counts on nutrient agar at two incubation temperatures, 20° C and 35° C, provided a useful indication of the organic pollution load in water.

2. The determination of fecal contamination, and hence the possible presence of pathogens, with the help of suitable **indicator organisms.** Indicator organisms are normally nonpathogenic, always occur in large quantities in feces, and are relatively easy to detect as compared to waterborne pathogen detection. The diagnosis of the latter is usually more complicated and time-consuming, and thus less suited for routine investigations. Assuming that in cold surface waters the pathogens are dying off faster than the indicator organisms employed, the absence of the latter or their presence in very low numbers guaranteed, in most cases, the absence of pathogens. Recently, however, better techniques and procedures have been developed for the detection of pathogenic bacteria and viruses. In certain cases, these techniques have indicated the presence of pathogens in the absence of indicator organisms hitherto relied upon, showing increased resistance of pathogens to the aqueous environment. Similar findings have been found with respect to resistance to chlorination. Such results suggest that some of the principles and methods in conventional water examination are of questionable value.

Nevertheless, conventional methods for detection of fecal contamination are still widely employed. Some will be described here. In general, these procedures are the same ones described by the American Public Health Association in *Standard Methods for the Examination of Water and Wastewater.*

The indicator organisms most widely used belong to the so-called **coliform group.** This includes all aerobic and facultative anaerobic, gram-negative, non-spore-forming, rod-shaped bacteria that ferment lactose with gas formation within 48 hours at 37° C, and comprises *Escherichia coli* (10^6 to 10^9 cells/g of feces) together with a number of closely related organisms (see exercise 25). **Noncoliforms** that are sometimes employed, primarily for confirmation, include *Streptococcus faecalis*, some related species, and in Great Britain, *Clostridium perfringens*, which is also called *C. welchii.*

The presence of *E. coli* in water from sources such as reservoirs suggests that chlorination is inadequate. Current standards for drinking water state that it should be free of coliforms and contain no more than 10 other microorganisms per ml.

Two of the most important methods applied to detect coliform organisms are the multiple-tube fermentation technique and the membrane filter technique.

Multiple-Tube Fermentation Technique

This technique employs three consecutive tests: first a presumptive test; if the first test is positive, then a confirmed test; and finally a completed test (figure 34.4 presents a pictorial description of these tests).

Presumptive Test This test, a specific enrichment procedure for **coliform bacteria,** is conducted in fermentation tubes filled with a **selective growth medium** (MacConkey lactose broth) and inverted Durham vials for detection of fermentation gas.

The main selective factors found in the medium are lactose, sometimes a **surfactant** such as Na-lauryl sulfate or Na-taurocholate (bile salt), and often a pH indicator dye for facilitating detection of acid production, such as bromcresol purple or brilliant green. The selective action of lactose occurs because many bacteria cannot ferment this sugar, whereas coliform bacteria and several other bacterial types can ferment it. The surfactant and dye do not inhibit coliform bacteria, whereas many other bacteria, such as the spore formers, are inhibited. The original surfactant used by MacConkey

Figure 34.4 Standard methods procedure for the examination of water and wastewater and for use in determining most probable number (MPN).

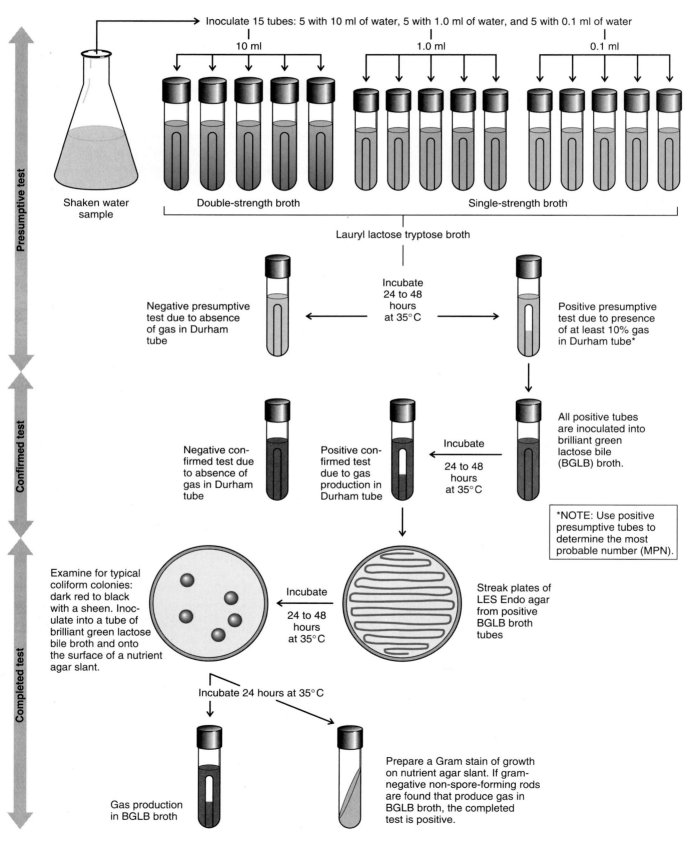

Inoculate 15 tubes: 5 with 10 ml of water, 5 with 1.0 ml of water, and 5 with 0.1 ml of water

10 ml — 1.0 ml — 0.1 ml

Shaken water sample

Double-strength broth — Single-strength broth

Lauryl lactose tryptose broth

Presumptive test

Negative presumptive test due to absence of gas in Durham tube

Incubate 24 to 48 hours at 35°C

Positive presumptive test due to presence of at least 10% gas in Durham tube*

Confirmed test

Negative confirmed test due to absence of gas in Durham tube

Positive confirmed test due to gas production in Durham tube

Incubate 24 to 48 hours at 35°C

All positive tubes are inoculated into brilliant green lactose bile (BGLB) broth.

*NOTE: Use positive presumptive tubes to determine the most probable number (MPN).

Completed test

Examine for typical coliform colonies: dark red to black with a sheen. Inoculate into a tube of brilliant green lactose bile broth and onto the surface of a nutrient agar slant.

Incubate 24 to 48 hours at 35°C

Streak plates of LES Endo agar from positive BGLB broth tubes

Incubate 24 hours at 35°C

Gas production in BGLB broth

Prepare a Gram stain of growth on nutrient agar slant. If gram-negative non-spore-forming rods are found that produce gas in BGLB broth, the completed test is positive.

was bile salt because *E. coli* is adapted to growing in the colon, whereas nonintestinal bacteria generally are not.

The formation of 10% gas or more in the Durham tube within 24–48 hours, together with turbidity in the growth medium constitutes a positive presumptive test for coliform bacteria, and hence for the possibility of fecal pollution. The test is presumptive only, because under these conditions several other types of bacteria can produce similar results.

The presumptive test also enables quantitation of the bacteria present in the water sample. The test, described as the most probable number test (MPN), is also useful for counting bacteria that reluctantly form colonies on agar plates or membrane filters, but grow readily in liquid media. In principle, the water sample is diluted so that some of the broth tubes contain a single bacterial cell. After incubation, some broth tubes show growth with gas, whereas others do not. The total viable count is then determined by counting the portion of positive tubes and referring this data to a statistical MPN table used for calculating the total viable bacterial count (see Procedure, table 34.1).

Confirmed Test This test serves to confirm the presence of coliform bacteria when either a positive or doubtful presumptive test is obtained. A loopful of growth from such a presumptive tube is transferred into a tube of brilliant green bile 2% broth and incubated at 35° C for 48 hours. This is a selective medium for detecting coliform bacteria in water, dairy, and other food products. In order to do so, the correct concentration of the dye (brilliant green) and bile must be present. If it is too concentrated, coliform growth can also be inhibited. Bile is naturally found in the intestine where it serves a similar purpose, encouraging growth of coliform bacteria while discouraging growth of other bacteria. A final selective agent in the medium is lactose. The broth tube also contains a Durham tube to detect gas production. The presence of gas in the Durham tube after incubating for 24 to 48 hours constitutes a positive confirmed test.

Completed Test This test helps to further confirm doubtful and, if desired, positive confirmed test results. The test is in two parts:

1. A plate of LES Endo agar (or Levine's EMB agar) is streaked with a loopful of growth from a positive confirmed tube, and incubated at 35° C for 18–24 hours. Typical coliform bacteria (*E. coli* and *Enterobacter aerogenes*) exhibit good growth on this medium and form red to black colonies with a sheen. *Salmonella typhi* exhibits good growth but the colonies are colorless. *S. aureus* growth is inhibited altogether.

2. Next a typical coliform colony from an LES Endo agar plate is inoculated into a tube of brilliant green bile broth and on the surface of a nutrient agar slant. They are then incubated at 35° C for 24 hours. A Gram stain is then prepared from growth present on the nutrient agar slant. The presence of gas in the brilliant green bile broth tube and gram-negative non-spore-forming rods constitutes a positive completed test for the presence of coliform bacteria, which, in turn, infers possible contamination of the water sample with fecal matter.

Membrane Filter Technique

For this technique, a known volume of water sample (100 ml) or of its dilutions is filtered by suction through a sterile polycarbonate or nitrocellulose acetate membrane filter. The filter is very thin (150 μm), and has a pore diameter of 0.45 μm. The precision manufacture of the filter is such that bacteria larger than 0.47 μm cannot pass through it. Filters with printed grid patterns are available for facilitating colony counting.

Once the water is filtered, the filter disc is aseptically transferred to the surface of a wetted pad contained in a Petri dish. The pad is soaked with Endo broth MF on which coliform colonies will appear to be pink to dark red in color and possess a golden green metallic sheen (see color plate 26). After incubation at 35° C for 24 hours, the filter disc is examined for characteristic coliform colonies and their number. From this number, one can calculate the total number of coliform bacteria present in the original water sample. For additional confirmation, the procedure for the completed multiple tube fermentation test can be applied.

The membrane filter method yields accurate results if the coliform colony count is in the range of 30 to 300 organisms per filter disc. Unknown samples require that various dilutions be examined. Turbidity of the sample constitutes a serious obstacle in cases where dilutions, yielding coliform counts within the desired range, contain sufficient suspended matter to clog the filter before the required volume of water has passed through.

The advantages of the membrane filter technique over the multiple-tube fermentation test are: (1) better reproducibility of results; (2) greater sensitivity, because larger amounts of water can be used; and (3) shorter time (one-quarter the time) for obtaining results. This method has been recognized by the United States Public Health Service for detection of coliforms in water.

Definitions

Coliform bacteria. A collective term for bacteria that inhabit the colon and are able to ferment lactose.

Indicator organism. A nonpathogenic organism whose presence when detected in water or sewage serves as an indicator of possible pollution with pathogens.

Selective growth medium. A growth medium that contains substances that inhibit the growth of certain organisms but not others.

Surfactant. A surface-active agent that forms a water-soluble common boundary between two substances. Examples include detergents and wetting agents.

Objectives

1. To introduce you to the use of a multiple-tube fermentation technique for detecting the presence and number of coliform pollution indicator organisms present in water samples.
2. To introduce you to the use of an alternate method, the membrane filter technique, for detecting the presence and number of coliform bacteria in water samples.

References

McKinney, R. E. *Microbiology for sanitary engineers*, New York: McGraw-Hill Book Co., 1972.

Nester et al. *Microbiology: A human perspective.* Chapters 5, 120; 11, 228, 243, 248–250; 31, 699–701; 32, 703–715.

Standard Methods for the Examination of Water and Wastewater, 18th ed. The American Water Works Association, 6666 West Quincy Ave., Denver, Co 80235.

Materials

Multiple-Tube Fermentation Technique (per student)

Test tubes, 50 ml, containing 10 ml of double strength lauryl sulfate (lauryl lactose tryptose) broth, 5

Small test tubes containing 10 ml of single strength lauryl sulfate broth, 10

Sterile 10-ml pipet, 1

Sterile 1-ml pipet calibrated in 0.1 ml units, 1

Brilliant green bile 2% broth, 2 tubes

LES Endo agar plate, 1

Nutrient agar slant, 1

Sterile 100-ml screw cap bottle for collecting water sample, 1

Membrane Filter Technique (demonstration)*

A 1-liter side-arm Erlenmeyer flask, 1

Sterile membrane filter holder assembly, two parts wrapped separately (see figure 34.5, frames 2 and 3), 1 unit

A metal clamp for clamping filter funnel to filter base

Sterile membrane filters, 47 mm diameter, 0.45 μm pore size

Forceps, 1 pair

Sterile 50-mm diameter Petri dishes, 3

Absorbent filter pads, 3

*Note: For student participation, see Appendix 5, Notes to Instructor, concerning availability of an inexpensive sterile membrane filtration system.

Tube containing 10 ml of sterile Endo MF broth, 1

Sterile 90-ml water blanks, 2

Erlenmeyer flasks containing 25 ml of sterile water, 6

Sterile 10-ml pipets, 2

Vacuum pump or Venturi vacuum system

Water sample for coliform analysis, 100 ml

Procedure

Multiple-Tube Fermentation Technique

Note: If desired, your instructor may ask you to bring a 50- to 100-ml sample of water from home, a nearby stream, a lake, or some other location for analysis. When taking a tap sample, the orifice of the water tap should be flamed before being opened. After opening, allow water to run for 5 to 10 minutes with the tap in the same position to prevent loosening of bacteria from inside the tap. Next, using aseptic technique, open a sterile bottle (obtained from the instructor beforehand) and collect a sample. If the sample cannot be examined within 1–2 hours, keep refrigerated until ready for use.

First Session (Presumptive Test)

1. Shake water sample. Aseptically pipet 10-ml portions of the sample into each of the five large tubes containing 10-ml aliquots of double-strength lauryl sulfate broth. Next, with a 1-ml pipet, transfer 1-ml portions of the water sample into five of the smaller tubes, and 0.1-ml portions into the remaining five small tubes of lauryl sulfate broth. Be sure to label the tubes.
2. Incubate the test tubes for 48 hours at 37° C.

Second and Third Sessions (Presumptive and Confirmed Tests)

1. Observe the tubes after 24 hours for gas production by gently shaking the tubes. If after shaking gas is not evident in the Durham tube, reincubate the tube for an additional 24 hours. Record any positive results for gas production in table 34.2 of the Laboratory Report.
2. Observe the tubes for gas production and turbidity after 48 hours of incubation. If neither gas nor turbidity are present in any of the tubes, the test is negative. If turbidity is present but no gas, the test may be doubtful since growth without gas may mean inhibition of coliform bacteria by noncoliform bacteria with shorter generation times. If at least 10% of the tube is filled with gas, the test is positive for coliform bacteria. Record your results in table 34.2 of the Laboratory Report.

Table 34.1   MPN Index and 95% Confidence Limits for Various Combinations of Positive Results When Five Tubes Are Used Per Dilution (10 mL, 1.0 mL, 0.1 mL)

Combination of Positives	MPN Index/ 100 mL	95% Confidence Limits		Combination of Positives	MPN Index/ 100 mL	95% Confidence Limits	
		Lower	Upper			Lower	Upper
0-0-0	<2	—	—	4-3-0	27	12	67
0-0-1	3	1.0	10	4-3-1	33	15	77
0-1-0	3	1.0	10	4-4-0	34	16	80
0-2-0	4	1.0	13	5-0-0	23	9.0	86
1-0-0	2	1.0	11	5-0-1	30	10	110
1-0-1	4	1.0	15	5-0-2	40	20	140
1-1-0	4	1.0	15	5-1-0	30	10	120
1-1-1	6	2.0	18	5-1-1	50	10	150
1-2-0	6	2.0	18	5-1-2	60	30	180
2-0-0	4	1.0	17	5-2-0	50	20	170
2-0-1	7	2.0	20	5-2-1	70	30	210
2-1-0	7	2.0	21	5-2-2	90	40	250
2-1-1	9	3.0	24	5-3-0	80	30	250
2-2-0	9	3.0	25	5-3-1	110	40	300
2-3-0	12	5.0	29	5-3-2	140	60	360
3-0-0	8	3.0	24	5-3-3	170	80	410
3-0-1	11	4.0	29	5-4-0	130	50	390
3-1-0	11	4.0	29	5-4-1	170	70	480
3-1-1	14	6.0	35	5-4-2	220	100	580
3-2-0	14	6.0	35	5-4-3	280	120	690
3-2-3	17	7.0	40	5-4-4	350	160	820
4-0-0	13	5.0	38	5-5-0	240	100	940
4-0-1	17	7.0	45	5-5-1	300	100	1300
4-1-0	17	7.0	46	5-5-2	500	200	2000
4-1-1	21	9.0	55	5-5-3	900	300	2900
4-1-2	26	12	63	5-5-4	1600	600	5300
4-2-0	22	9.0	56	5-5-5	≥1600	—	—
4-2-1	26	12	65				

From *Standard Methods for the Examination of Water and Wastewater,* 18th edition. Copyright 1992 by the American Public Health Association, the American Water Works Association, and the Water Environment Federation. Reprinted with permission.

3. MPN determination. Using your fermentation gas results in table 34.2, determine the number of tubes from each set containing 10% gas or more. Determine the MPN by consulting table 34.1; for example, if you had gas in two of the first five tubes, in two of the second five tubes, and none in the third three tubes, your test readout would be 2-2-0. Table 34.1 shows that the MPN for this readout would be 9. Thus, your water sample would contain nine organisms per 100-ml water with a 95% statistical probability of there being between three and twenty-five organisms.

Note: If your readout for the series is 0-0-0, it means that the MPN is less than two organisms per 100 ml of water. Also, if the readout is 5-5-5 it means the MPN is greater than 1,600 organisms/100 ml water. In the latter instance, what procedural modification would be required to obtain a more significant result? Report your answer in question 10 of the Questions section in the Laboratory Report.

4. The Confirmed Test should be administered to all tubes demonstrating either a positive or doubtful presumptive test. Inoculate a loopful

of growth from each tube showing gas or dense turbidity into a tube of brilliant green lactose bile 2% broth. Incubate the tube(s) at 37° C for 24 to 48 hours.

Note: For expediency, your instructor may wish you to inoculate only one tube. If so, for the inoculum use the tube of lauryl sulfate broth testing positive with the least inoculum of water.

Fourth Session (Confirmed and Completed Tests)

1. Examine the brilliant green lactose bile 2% tube(s) for gas production. Record your findings in the Confirmed test section of the Laboratory Report.
2. Streak a loopful of growth from a positive tube of brilliant green lactose bile 2% broth on the surface of a plate containing LES Endo agar. Incubate at 37° C for 24 hours.

Fifth Session (Completed Test)

1. Examine the LES Endo agar plate(s) for the presence of typical coliform colonies (dark red to black with a sheen). Record your findings in the Completed test section of the Laboratory Report.
2. With a loop, streak a nutrient agar slant with growth obtained from a typical coliform colony found on the LES Endo agar plate. Also inoculate a tube of brilliant green lactose bile 2% broth with growth from the same colony. Incubate the tubes at 37° C for 24 hours.

Sixth Session (Completed Test)

1. Examine the brilliant green lactose bile 2% broth tube for gas production. Record your result in the Completed test section of the Laboratory Report.
2. Prepare a Gram stain of some of the growth present on the nutrient agar slant. Examine the slide for the presence of gram-negative, non-spore-forming rods. Record your results in the Completed test section of the Laboratory Report. The presence of gas and of gram-negative, non-spore-forming rods constitutes a positive completed coliform test.

Membrane Filter Technique

First Session

1. Shake the water sample. Prepare two dilutions by transferring successive 10-ml aliquots into 90-ml blanks of sterile water (10^{-1} and 10^{-2} dilutions).

   Note: Reshake prepared dilutions before using.

2. Assemble the filter holder apparatus as follows (figure 34.5):
   a. Using aseptic technique, unwrap the lower portion of the filter base, and insert a rubber stopper.
   b. Insert the base in the neck of the 1-liter side-arm Erlenmeyer flask (figure 34.5, frame 2).
   c. With sterile forceps (sterilize by dipping in alcohol and flaming them in the flame of the Bunsen burner), transfer a sterile membrane filter onto the sintered glass or plastic surface of the filter holder base (figure 34.5, frames 1 and 2). Make certain the membrane filter is placed with the ruled side up.
   d. Aseptically remove the covered filter funnel from the butcher paper, and place the lower surface on top of the membrane filter. Clamp the filter funnel to the filter base with the clamp provided with the filter holder assembly (figure 34.5, frame 3).
3. Prepare 3 plates of Endo medium by adding 2-ml aliquots of the tubed broth to sterile absorbent pads previously placed aseptically with sterile tweezers on the bottom of the three Petri dishes (figure 34.5, frames 4 and 5).
4. Remove the aluminum filter cover and pour the highest water dilution (10^{-2}) into the funnel (figure 34.5, frame 6). Assist the filtration process by turning on the vacuum pump or Venturi system.
5. Rinse the funnel walls with two 25-ml aliquots of sterile water.
6. Turn off (break) the vacuum and remove the filter holder funnel. Using aseptic technique and sterile tweezers, transfer the filter

Figure 34.5 Analysis of water for fecal contamination. Cellulose acetate membrane filter method. (*1*) Sterile membrane filter (0.45 µm) with grid for counting is handled with sterile forceps. (*2*) The sterile membrane filter is placed on filter holder base with grid side up. (*3*) The apparatus is assembled. (*4*) Sterile absorbent pads are aseptically placed in the bottom of three sterile Petri dishes. (*5*) Each absorbent pad is saturated with 2.0 ml of Endo broth. (*6*) A portion of well-mixed water sample is poured into assembled funnel and filtered by vacuum. (*7*) Membrane filter is carefully removed with sterile forceps after disassembling the funnel. (*8*) Membrane filter is centered on the surface of the Endo-soaked absorbent pad (grid side up) by use of a slight rolling motion. (*9*) After incubation, the number of colonies on the filter is counted. The number of colonies on the filter reflects the number of coliform bacteria present in the original sample.

immediately to the previously prepared Petri dish (figure 34.5, frames 7 and 8). Using a slight rolling motion, center the filter, grid side up, on the medium-soaked absorbent pad. Take care not to trap air under the filter as this will prevent nutrient media from reaching all of the membrane surface (figure 34.5, frame 9).

7. Reassemble the filter apparatus with a new membrane filter, and repeat the filtration process first with the 10^{-1} water sample, and finally with a 100-ml aliquot of the undiluted water sample.

8. Label and invert the Petri dishes to prevent any condensate from falling on the filter surface during incubation. Incubate plates for 24 hours at 37° C.

Second Session

1. Count the number of coliform type bacteria by using either the low power of the microscope or a dissecting microscope. Count only those colonies that exhibit a pink to dark red center with or without a distinct golden green metallic sheen (see color plate 26).

2. Record the number of colonies found in each of the three dilutions in table 34.3 of the Laboratory Report.

NOTES:

Name _____ Date _____ Section _____

EXERCISE

Laboratory Report: Bacteriological Examination of Water: Multiple-Tube Fermentation and Membrane Filter Techniques

Results

1. Multiple-Tube Fermentation Technique
   a. Record the results of the presumptive test in table 34.2.

Table 34.2 Presumptive Test for the Presence or Absence of Gas and Turbidity in Multiple-Tube Fermentation Media

Water Sample Size (ml)	Presence of Gas and Turbidity*				
	Tube #1	Tube #2	Tube #3	Tube #4	Tube #5
10					
1					
0.1					

*Use a (+) sign to indicate gas and a circle (O) around the plus sign to indicate turbidity

   b. Determine the MPN below:

Test readout: _____   MPN: _____   95% Confidence Limits: _____

   c. Confirmed test results (gas production in brilliant green lactose bile 2% broth):

Sample Number   Gas (+ or –)
                  24 hours   48 hours

   d. Appearance of colonies on LES Endo agar:

   e. Completed test results:

Sample Number   Gas (+ or –)   Gram Stain Reaction

2. Membrane Filter Technique

Table 34.3   Number of Coliform Colonies Present in Various Dilutions of the Water Sample

Undiluted Sample	10⁻¹ Dilution	10⁻² Dilution

    a. Calculate the number of coliform colonies/ml present in the original water sample (show your calculations):

Questions

1. What nutritional means might be used to speed up the growth of the coliform organisms using the membrane filter technique?

2. Describe two other applications of the membrane filter technique.

3. Why not test for pathogens such as *Salmonella* directly rather than use an indicator organism such as the coliform bacteria?

4. Why does a positive presumptive test not necessarily indicate that the water is unsafe for drinking?

5. List three organisms that are apt to give a positive presumptive test.

6. Describe the purpose of lactose and Endo agar in these tests.

7. What are some limitations of the membrane filter technique?

8. Define the term *coliform*.

9. Briefly explain what is meant by presumptive, confirmed, and completed tests in water analysis.

10. See Note, MPN determination, on page 306 for the question.

APPENDIX

1

List of Microbial Cultures, Methods for Storage, and Sources

Exercise 1

Demonstration cultures of bacteria and molds (optional)

Exercise 2

Baker's yeast, grocery store, dried or cake

Exercise 3

Baker's yeast, grocery store, dried or cake
Staphylococcus epidermidis
Spirillum volutans
Bacillus cereus
Pseudomonas aeruginosa

Exercise 4

Bacillus cereus or *Bacillus subtilis*
Staphylococcus epidermidis
Enterococcus faecalis
Escherichia coli

Exercise 5

Escherichia coli
Bacillus subtilis
Bacillus cereus
Staphylococcus epidermidis
Enterococcus faecalis
Mycobacterium smegmatis
Klebsiella pneumoniae
Micrococcus luteus
Spirillum

Exercise 6

Staphylococcus epidermidis
Micrococcus luteus

Exercise 7

Escherichia coli
Staphylococcus epidermidis
Pseudomonas aeruginosa
Enterobacter aerogenes

Exercise 8

Escherichia coli K12

Exercise 9

Clostridium sporogenes
Escherichia coli
Micrococcus luteus

Exercise 10

Escherichia coli

Exercise 11

Staphylococcus epidermidis
Escherichia coli
Bacillus cereus
Bacillus subtilis

Exercise 12

Escherichia coli
Bacillus cereus or *Bacillus subtilis* spores in saline

Exercise 13

Escherichia coli
Halobacterium salinarium
Saccharomyces cerevisiae
Micrococcus luteus

Exercise 14

Staphylococcus epidermidis
Escherichia coli
Pseudomonas aeruginosa
Mycobacterium smegmatis

Exercise 15

Escherichia coli K12

Exercise 16

Acinetobacter, streptomycin sensitive ATCC 33305
Acinetobacter, streptomycin resistant ATCC 33969

Exercise 17

Escherichia coli A F^+ str^S met^- thr^+ ATCC 29839
Escherichia coli B F^- str^R met^+ thr^- ATCC 29842

Exercise 18

Escherichia coli
Bacillus cereus

Exercise 19

Saccharomyces cerevisiae
Candida albicans

Exercise 20

Rhizopus nigricans
Aspergillis niger
Penicillium notatum or sp.

Exercise 21

Living *Paramecium caudatum* CBS L-2A
Living *Amoeba proteus* CBS L-1
Dugesia tigrina CBS L-229D

Exercise 22

Escherichia coli K12
Escherichia coli phage T4

Exercise 23

Optional for demonstration
Staphylococcus epidermidis
Micrococcus luteus
Staphylococcus aureus
Propionibacterium acnes

Exercise 24

Streptococcus pyrogenes Group A β-hemolytic
α-hemolytic *Streptococcus* sp.

Exercise 25

Escherichia coli
Enterobacter aerogenes
Klebsiella pneumoniae
Pseudomonas aeruginosa
Proteus mirabilis

Exercise 26

Variable, but can include the following:
Staphylococcus epidermidis
Escherichia coli
Enterobacter aerogenes
Pseudomonas aeruginosa

Exercise 27

None

Exercise 28

Staphylococcus aureus
Escherichia coli
Saccharomyces cerevisiae

Exercise 29

None

Exercise 30

Streptococcus pyogenes (Lancefield Group A)
Streptococcus pneumoniae (Lancefield Group B)

Exercise 31

Streptococcus pyogenes (Lancefield Group A)
Streptococcus mutans or other viridans streptococcus

Exercise 32

None

Exercise 33

Staphylococcus aureus ATCC 25923 (see exercise 14)

Exercise 34

None

Note: ATCC has Preceptrol strains available at a lower cost for teaching.
ATCC strains are necessary when indicated.

Culture Preservation and Storage

Bacterial cultures are critical to the success of microbiology experiments. Therefore, a major concern for microbiology instructors is maintaining and storing the culture collection between courses. The following list of methods includes those currently in use in our department. Unfortunately, the most dependable methods seem to require the most elaborate and expensive equipment. However, there are some simple methods that are very useful and reliable for teaching laboratories. It is a good idea to store an important culture several different ways, so that if a freezer fails or an agar slant dries up, there are back-up cultures.

Note: A standard tube refers to a 15-cm tube. An agar deep contains 12 ml of agar in a standard tube.

1. Storage in refrigerator
   In our labs, cultures (in addition to long storage) are carried on nutrient agar slants in screw cap tubes and transferred every three months. Nutrient agar is chosen because, as a rule, the less rich the medium, the longer the culture will remain viable. Not all organisms will grow on the relatively lean nutrient agar, so these must be carried on a richer medium such as trypticase soy agar or blood agar. The cultures are refrigerated and most cultures are viable at least three months. Note the following exceptions:
   a. α- and β-hemolytic *Streptococcus* are transferred monthly on blood agar slants.
   b. *Spirillum* can be stored in a dilute medium called cytophaga agar. Prepare agar deeps of cytophaga agar in standard screw cap tubes. Inoculate the *Spirillum* just below the surface of the agar and incubate 2–3 days at room temperature until a band of growth appears a few millimeters below the surface of the agar (they are microaerophilic). Store in the refrigerator. These will remain viable at least three months and usually much longer.

2. Storage at room temperature
   Prepare agar deeps of 1% peptone plus 1% agar in standard screw cap tubes. Touch loop to several average-sized colonies on an overnight streak plate and stab agar to the bottom of the tube. Inoculating just one colony might result in selecting a mutant. Incubate at the normal temperature overnight. Tighten cap, wrap cap with parafilm, and store at room temperature. *E. coli* cultures have remained viable for at least eight years and *Acinetobacter* at least one year.

   Many strains of *Clostridium* can be grown in thioglycollate broth and stored at room temperature. Be sure to boil the thioglycollate broth for several minutes to drive out the oxygen and then cool before inoculating.

3. Storage in freezer of –20° C in small vials
   a. TS broth + 15% glycerol or
   b. TS broth and fetal calf serum mixed 1:1
      Two ml of broth are placed in screw cap vials (4 ml capacity) and colonies of fresh cultures are suspended in the broth. Tighten screw caps and freeze in standard freezer. Do not store in a frostfree freezer because it has a thaw and freeze cycle that reduces viability. To grow, thaw at room temperature and streak on an agar plate.

      Broth containing 15% glycerol is commonly used for freezing cultures, and is inexpensive and readily available. Some research laboratories use fetal calf serum broth, but it is quite expensive.

4. Storage at $-180°$ C
   Prepare as in 3 but store at the lower temperature.
5. Freeze-drying
   This method of preserving cultures is the most reliable for long-term storage (decades), but is also the most technically demanding. Special apparatus for obtaining a vacuum and sealing tubes must be used as the cultures are frozen. It is an excellent method, but not always available.

Storing Virus

1. Viruses are stored in the refrigerator in tightly capped tubes or bottles. If the broth is contaminated with bacteria, add a few drops of chloroform. This will kill the bacteria, but not harm most viruses.
2. To renew phage stocks of lytic phage try one of the following: (a student could try these as a special project)
   a. Grow the host organism to about 1×10^8 cells/ml—5 ml might be a good amount. Infect with a multiplicity of infection of at least 4 (or in other words, add about 4 virus particles to the broth for every bacterial cell) to ensure that each bacterium is infected. Shake the culture at $37°$ C and watch for lysis (clearing) of the culture—about 2 hours in *E. coli*. (This can be difficult because sometimes not all the cells lyse and the culture can start growing again so that it never clears.) Add chloroform (1%), let stand 15 minutes, and store in refrigerator.
   b. Dilute the phage to about 1×10^4, 1×10^5, and 1×10^6 phage/ml. Grow the host organism to about $OD_{600} = 0.4$ or late log. Mix 0.1 ml of phage and 0.1 ml of the host bacterial cells in 4.0 ml melted, cooled, soft overlay agar and pour on an agar base plate. After 12 hours incubation choose the plate with confluent lysis in which the plaques are just touching each other. Pour 5.0 ml of TS broth over plate and let incubate for several hours so that phage diffuse out of the plaques into the broth. Remove broth with a pipet, add a few drops of chloroform, and store in refrigerator. This is a modified procedure from Sambrook et al., *Molecular Cloning*, 2d edition, Cold Spring Harbor Laboratory Press, 1989.

Catalog Sources of Cultures, Media, Equipment, and Biochemicals

American Type Culture Collection
12301 Parklawn Drive
Rockville, Maryland 20852–1776

Catalog: *American Type Culture Collection*, 18th ed., 1992

Telephone USA/Canada 1–800–638–6597
Outside USA/Canada 1–301–881–2600
FAX 1–301–231–5826
Technical information telephone 1–301–881–2600 (1:30–4:30 P.M. ET)
   ATCC offers at a reduced rate cultures which are especially useful for teaching. These are called Preceptrol cultures. An account must be set up before ordering. The complete set of ATCC catalogs are available online. Call BIOSIS at 1–800–523–4806 for information.

Becton Dickinson Microbiology Systems
P.O. Box 243
Cockeysville, Maryland 21030–0243

Catalog: *Consumer Catalog Becton Dickinson Microbiology Systems*

   A source of dehydrated and prepared microbiological media. BBL products can be obtained only through a distributor.
   Telephone 1–410–584–7154 or write the Consumer Service at the above address for a Consumer Catalog and a list of local distributors.
   The Manual of BBL Products and Laboratory Procedures is also available and could be very useful, but it costs $25.00

Carolina Biological Supply Company
2700 York Road
Burlington, North Carolina 27215

Catalog: *Biology/Science Materials*

Eastern half of U.S.
Telephone 1–800–334–5551
FAX 1–919–584–3399
Western half of U.S.
Telephone 1–800–547–1733
FAX 1–503–656–4208

This catalog is oriented to high schools, but their phage and bacterial cultures are very reasonably priced. Excellent source for prepared slides.

Difco Laboratories
Detroit, Michigan 48232

Manual: *Difco Manual*, 10th ed.

Catalog: *Difco Product Catalog and Reference Guide*

Telephone 1–800–521–0851 for the *Difco Product Catalog*, information, and local distributors.

An old reliable source of microbiological media. Difco sells only through dealers. The manual contains much helpful information, but there may be a charge for it.

Midwest Culture Service
1924 North Seventh Street
Terre Haute, Indiana 47804

Telephone 1–812–232–6785

Catalog of Cultures
A fairly good selection of modestly priced cultures.

Sigma Chemical Company
P.O. Box 14508
St. Louis, Missouri 63178–9916

Catalog: *Biochemicals, Organic Compounds for Research, and Diagnostic Reagents*

USA/Canada Telephone 1–800–325–3010
USA/Canada FAX 1–800–325–5052
Outside USA/Canada FAX 1–314–771–5757

Wide variety of biochemical compounds and reagents.

Unipath (Oxoid)
P.O. Box 691
Ogdensburg, New York 13669

VWR Scientific
P.O. Box 13645
Philadelphia, Pennsylvania 19101–9711

VWR has a catalog of equipment of all kinds as well as media and reagents. Write for information on current catalogs and the sales office in your area.

APPENDIX 2

Stains and Biochemical Test Reagents

Exercise 3

Acidified methylene blue

Exercise 4

Staining reagents in 250-ml plastic bottles
Methylene blue
Crystal violet
95% ethanol or acetone/95% ethanol 1:1
Safranin
India ink

Exercise 5

Crystal violet
Gram's iodine
95% alcohol or acetone/alcohol
Safranin
Carbolfuchsin
Methylene blue
Acid alcohol
India ink
Malachite green

Exercise 12

Sterile saline

Exercise 14

3% H_2O_2
0.3% hexachlorophene
70% Ethanol
Antiseptic mouthwash
Other antiseptics
Vials of antibiotic discs
  Penicillin, 10 µg; streptomycin, 10 µg;
  tetracycline, 30 µg; chloramphenicol, 30 µg;
  nystatin, 100 units; sulfanilamide (or another
  sulfonamide), 300 µg; erythromycin, 15 µg.

Exercise 15

Streptomycin

Exercise 16

Sodium dodecyl sulfate in 10× saline citrate
DNase
Streptomycin

Exercise 18

ONPG
Gram's iodine

Exercise 19

Methylene blue

Exercise 21

Planaslo solution

Exercise 22

Chloroform (optional)

Exercise 23

Gram-stain reagents
95% ethanol
Dehydrated plasma
Sterile saline

Exercise 24

3% H_2O_2 in dropper bottles
Sterile saline
Oxidase reagent

Exercises 25 and 26

Gram-stain reagents
Kovacs reagent
Methyl red
Voges-Proskauer reagents A and B
Oxidase reagent
H_2O_2
India ink
Malachite green

Exercise 27

Wright's stain
Phosphate buffer, pH 6.8

Exercise 28

Lysozyme chloride

Exercise 29

Nontreponemal cardiolipin antigen, Difco catalog
 #2405-46-3 (6 vials each containing 3 ml)
Nontreponemal antigen reactive antiserum,
 Difco catalog #3516-32-2 (3 vials containing
 positive, weakly reactive, and nonreactive
 control)
Isotonic saline in distilled water
Proteus OX-2 antigen for slide test, 5 ml, Difco Cat.
 # 2243-56-6
Proteus OX-2 antigen for tube test, 25 ml, Difco
 Cat. # 2248-65-0
Proteus OX-2 antiserum, for slide or tube test, 3 ml,
 Difco Cat. #2245-47-6
Proteus OX-19 antigen is available in either
 5-ml (slide test, Difco Cat. #2234-56-7) or
 25-ml (tube test, Difco Cat. #2247-65-1)
 quantities. The positive polyvalent control
 serum is available in 5-ml quantities (Difco
 Cat. #3238-56-1).

Exercise 30

Gram-stain reagents
BBL Strep Grouping Kit

Exercise 31

Abbott TestPak Strep A test kit

Exercise 32

Meridian immunodiffusion test kit

Exercise 33

Dehydrated plasma
3% H_2O_2
Antibiotic discs

Exercise 34

Gram-stain reagents

Stains and Biochemical Test Reagents: Formulae and Sources

Acid Alcohol for Acid-Fast Stain

ethanol (95%)	97.0 ml
concentrated HCl	3.0 ml

Add acid to alcohol. Do not breathe acid fumes.

Carbolfuchsin Stain

basic fuchsin	0.3 g
ethanol (95%)	10.0 ml
phenol	5.0 ml
distilled water	100.0 ml

Dissolve the basic fuchsin in ethanol. Add the
phenol to the water. Mix the solutions together.
Allow to stand for several days. Filter before use.

DNase (Deoxyribonuclease)

Sigma DN-25 or any crude grade.
Add a few visible crystals to 5 ml sterile water.
Hold in ice bucket when in solution.

Gram-Stain Reagents

These can be purchased premixed from Sigma.

Crystal violet (Hucker Modification)

Solution A: crystal violet (gentian violet)	40.0 g
ethanol (95%)	400.0 ml
Solution B: ammonium oxalate	1.0 g
distilled water	1600.0 ml

Dissolve the crystal violet in the alcohol and let stand overnight. Strain through filter paper. Add solution B. Mix to dissolve completely.

Gram's Iodine

iodine	1.0 g
Kl	2.0 g
distilled water	300.0 ml

Dissolve the Kl in 2 ml of water. Add the iodine and dissolve. Then add the 300 ml of water.

Decolorizer for Gram Stain

Use either 95% ethanol or ethanol/acetone 1:1.

Safranin O

Safranin	10.0 g
Distilled water	1000.0 ml

Hexachlorophene

Can be purchased at a pharmacy, but a prescription may be needed. Can also order from Sigma.

Hydrogen Peroxide (3%)

Purchase 3% H_2O_2 and put in dropper bottles. Refrigeration is recommended.

India Ink

Can be purchased at stationery store. Ink consisting of smaller particles is better than larger particles.

Kovacs Reagent

isoamyl alcohol (isopentyl)	75.0 ml
concentrated HCl	25.0 ml
p-dimethylaminobenzaldehyde	5.0 g

Mix the alcohol and the acid. Add the aldehyde and stir to dissolve. Store in plastic or dark glass bottle.

Malachite Green (5%)

malachite green	5.0 g
distilled water	95.0 ml

Mix, filter, and store in brown bottle.

Methyl Red

methyl red	0.1 g
ethanol (95%)	300.0 ml
distilled water	200.0 ml

Dissolve methyl red in alcohol and add water. Dispense into dropper bottles.

Methylene Blue (Loeffler's)

Solution A: methylene blue	0.3 g
95% ethanol	30.0 ml
Solution B: 0.01% (wt/vol)KOH 100.0 ml	
about 1 pellet KOH/1000 ml	

Mix solutions A and B together.

Methylene Blue (Acidified)

Solution A: methylene blue	0.02 g
distilled water	100.0 ml
Solution B: KH_2PO_4 (0.2 molar)	99.75 ml
Solution C: K_2HPO_4 (0.2 molar)	0.25 ml

Mix solutions A, B, and C together (pH must be 4.6).

ONPG

O-nitrophenyl β-D galactopyranoside	1.0 g (.0003 M)
distilled water	1000.0 ml

Oxidase Reagent

N,N-dimethyl-p-phenylene-diamine hydrochloride	1.0 g
distilled water	100.0 ml
pH 5.5	

This must be prepared on the day it is used. However, it can be frozen in small vials and thawed when needed, although the test seems to be a little more difficult to read.

Phosphate Buffer

Can be purchased from Sigma or made using the following formula:

$Na_2HPO_4 \cdot 2H_2O$	11.88 g
KH_2PO_4	9.08 g
distilled water	1000.0 ml

For pH 6.8: mix 6 parts Na_2HPO_4 with 4 parts KH_2PO_4.

Planaslo Solution

Can be purchased from Carolina Biological Supply.

Plasma (Dehydrated for Coagulase)

Can be purchased from Difco or BBL (Becton Dickinson).

Saline

NaCl	8.5 g
distilled water	1000 ml

Sodium Dodecyl Sulfate (SDS) in 10× Saline Citrate

sodium dodecyl sulfate	5.0 g
NaCl	87.7 g
sodium citrate	44.1 g
distilled water	1000 ml

No pH adjustment. Autoclave at 121° C for 15 minutes.

Streptomycin

Streptomycin sulfate Difco Catalog #S 6501. Prepare a stock solution by adding 1 g to 10 ml distilled water. Filter, sterilize, and freeze in convenient amounts.

Voges-Proskauer

Reagent A: alphanaphthol	5.0 g
ethyl alcohol (95%)	100 ml
	(bring to volume)
Reagent B: KOH	40 g
distilled water	100 ml
	(bring to volume)

Dispense each reagent in separate dropper bottles.

Wright's Stain (Prepared)

Can be obtained from Sigma.

APPENDIX

3

Culture Media

Exercise 1

TS agar plates

Exercise 3

Nutrient broth
Yeast fermentation broth containing glucose

Exercise 4

TS agar slants

Exercise 5

TS agar slants
Nutrient broth
Nutrient agar slant
Worfel-Ferguson agar (optional)

Exercise 6

TS agar plates
TS broth

Exercise 7

TS broth
TS agar plates
Glucose-mineral salts agar plates
EMB agar plates

Exercise 8

TS broth
TS agar deeps

Exercise 9

TS broth
TS agar slants
TS + 0.5% glucose agar deeps
Thioglycollate broth

Exercise 10

TS broth

Exercise 11

TS broth
Nutrient agar slant

Exercise 12

TS broth
TS agar plates
Sterile saline
Raw hamburger

Exercise 13

TYEG salts agar slants
Halobacterium agar slants (ATCC medium 213)
TYEG salts agar + NaCl plates (0.5, 5, 10, 20%)
TYEG salts agar + sucrose plates (0, 10, 25, 50%)

Exercise 14

Mueller-Hinton agar
TS broth

Exercise 15

TS broth
TS agar deeps
TS agar + streptomycin

Exercise 16

TSY agar plates
TSY agar plates + streptomycin

Exercise 17

Mineral salts + 0.5% glucose agar plates (mineral medium or MM)
TS agar plates

Exercise 18

Mineral salts + 0.2% glucose broth
Mineral salts + 0.2% lactose broth
Mineral salts + 0.2% glycerol broth
Nutrient + starch agar plates
Nutrient + starch + glucose agar plates

Exercise 19

Corn meal agar plates
Glucose–acetate yeast sporulation agar plates
Yeast fermentation broth containing either glucose,
lactose, or maltose and Durham tubes
Sabouraud's dextrose agar plates and slants
Carrot wedges
Raw egg white (nonsterile) or dehydrated serum
Sabouraud's dextrose broth

Exercise 20

Sabouraud's dextrose agar

Exercise 22

Tryptone agar base plates
Tryptone soft overlay agar
Tryptone broth

Exercise 23

TSY agar plates
Sterile saline
TSY + glucose agar deeps
TSY + glucose + brom cresol purple agar slants

Exercise 24

Blood agar plates
Sterile saline

Exercise 25

Glucose fermentation tubes
Lactose fermentation tubes
Sucrose fermentation tubes
Tryptone broth
Methyl red–Voges-Proskauer broth
Simmons citrate agar slants
Motility agar deeps
Sterile saline

Exercise 26

Media can include any media from exercises 23, 24,
and 25 plus the following:
TS agar plates
MacConkey agar plates
Mannitol salt agar plates
Nutrient agar slants
Urea slants

Exercise 28

Nutrient broth
Sabouraud's dextrose broth
Sabouraud's dextrose agar deeps
Nutrient agar deeps
Raw egg white (aseptically prepared)

Exercise 29

Isotonic saline in distilled water

Exercise 30

Blood agar plates
TS broth
Sterile physiological saline (0.85% NaCl)

Exercise 31

Blood agar plates

Exercise 33

TS broth
Mannitol salt agar plates with phenol red indicator
Blood agar plates
Mueller-Hinton agar plates

Exercise 34

Multiple-tube fermentation technique:
Lauryl sulfate broth (double and single strength)
Brilliant green lactose bile 2% broth
LES Endo agar plate
Nutrient agar slant
Membrane filtration technique:
Endo MF broth

Media for Storing Organisms

Nutrient agar slants
1% peptone agar deeps
Cytophaga agar deeps
TS broth + 15% glycerol

General Information

Most of the standard media listed can be purchased to make either broth or agar. It is frequently convenient to buy the broth form and add agar when needed.

TS is an abbreviation for tryptic (Difco) soy or trypticase (BBL) soy.

Dextrose, as it is frequently called in the food industry, is another name for glucose.

Percent indicates grams or ml/100 ml. For example, 10% glucose solution w/v (weight per volume) would be 10 grams of glucose in 100 ml of water; 3% v/v (volume per volume) H_2O_2 is 3 ml of H_2O_2 in 100 ml of water.

Molar solutions are the molecular weight of the substance in 1000 ml of water. The molecular weight or formula weight can usually be found on the bottle label. 0.1 M solution of glucose would be 0.1×180 (m.w. of glucose) in 1000 ml or 18 g of glucose dissolved and water added to 1000 ml.

Agar

Medium with agar can be made ahead of time, sterilized in flasks, and stored at room temperature. Using aluminum foil to cover flasks instead of cotton or caps is very convenient.

Agar can be melted on a hot plate, in a steamer, or in a microwave oven. Be careful that it doesn't burn or boil over.

Note the following preparations:
agar slants: 6 ml/standard tube (13 cm)
agar deeps: 12 ml/standard tube
agar plates: about 20 ml agar

Sterilization

Glassware can be sterilized by baking in an oven for 90 minutes at 350° F.

Disposable filters are very convenient for solutions. When filtering small amounts, it is especially convenient to use filters that fit on syringes.

Autoclaves have three cycles:
    regular—exhausts rapidly
    liquid—exhausts slowly so that liquids will
    not boil over
    wrapped material—exhausts along with a
    vacuum so that moist air is removed

When sterilizing tubes of broth with sugars, autoclave at 118° C for 12 minutes to avoid breaking down the sugars. Increase the time for larger volumes.

References for Further Information

Atlas, Ronald M. *Microbiological Media*. Boca Raton: CRC Press, 1993.
Merck Index. 11th ed. Rahway, New Jersey: Merck, 1989.
Meynell, G. G., and Elinor Meynell. *Theory and Practice in Experimental Bacteriology*. 2d ed. London: Cambridge University Press, 1975.

Blood Agar

Blood agar base (Difco)	40.0 g
Distilled water or deionized water	1000 ml

Dissolve and sterilize 15 minutes at 121° C. Cool to 45°–50° C. Add 5% sterile, defribinated sheep blood. Human blood (outdated from the blood bank) does not demonstrate good hemolysis. (You may be able to purchase these plates already prepared from a local firm or hospital.)

Brilliant Green Bile Broth (Brilliant Green Lactose Bile Broth)

Composition per liter:

Oxgall, dehydrated,	20.0 g
Lactose,	10.0 g
Pancreatic digest of gelatin,	10.0 g
Brilliant Green,	0.013 g

pH 7.2 ± 0.2 at 25° C

Source: This medium is available as a premixed powder from Difco Laboratories, BBL Microbiology Systems and Oxoid Unipath.
Preparation of medium: Add components to distilled/deionized water and bring volume to 1.0L. Mix thoroughly. Distribute into tubes containing inverted Durham tubes, in 10.0 mL amounts for

testing 1.0mL or less of sample. Autoclave for 12 min (not longer than 15 min) at 15 psi pressure–121° C. After sterilization, cool the broth rapidly. Medium is sensitive to light.

Carrot Wedges

Cut a carrot diagonally, place in a test tube with the diagonal edge at the top, and insert a stopper. Sterilize at 121° C for 15 minutes.

Corn Meal Agar

Corn meal	60.0 g
Agar	15.0 g
Tween 80*	10.0 g
Distilled water	1000 ml

Mix the corn meal to a smooth cream with water, simmer for one hour, filter through cheese cloth or coarse filter paper, add the agar and Tween 80 and heat till dissolved, make up to volume, and sterilize, preferably for 30 minutes at 10 pounds pressure. Commercial corn meal agar is also available.

*Tween 80 addition greatly stimulates chlamydospore formation.

Cytophaga Agar

Tryptone	0.5 g
Yeast extract	0.5 g
Beef extract	0.2 g
Sodium acetate	0.2 g
Tap water	1000 ml
Agar	4.0 g

pH 7.2 before sterilizing
pH 7.2–7.3 after sterilizing

Mix and sterilize at 121° C for 15 minutes. Dispense 10 ml/screw cap tube.

Endo Agar

Peptone	10.0 g
Lactose	100.0 g
KH_2PO_4	3.5 g
Sodium sulfate	2.5 g
Agar	10.0 g
Distilled water	

pH 7.5 (approx.)

Suspend in 1 liter distilled water. Add 4 ml of a 10% (weight/volume) alcoholic solution of basic fuchsin (95% ethyl alcohol). Bring to a boil to dissolve completely. Sterilize at 121° C for 15 minutes. Mix well before pouring.

Endo Broth

Same composition as Endo agar except agar is omitted.

Levine Eosin Methylene Blue (EMB) Agar

Peptone	10.0 g
Lactose	10.0 g
K_2HPO_4	2.0 g
Eosin Y	0.4 g
Methylene blue	0.065 g
Agar	15.0 g
Distilled water	1000 ml

Suspend in 1 liter of distilled water. Bring to boil to dissolve completely. Sterilize at 121° C for 15 minutes. Cool to 60° C and shake medium in order to oxidize the methylene blue (i.e., restore its blue color) and to suspend the precipitate which is an important part of the medium. Commercial EMB agar is also available.

Glucose–Acetate Yeast Sporulation Agar

Per liter: Glucose	1.0 g
Yeast extract	2.0 g
Sodium acetate (with 3 H_2O)	5.0 g
Agar	15.0 g

Adjust pH to 5.5. Sterilize 15 minutes at 121° C.

Halobacterium Medium (ATCC Medium 213)

NaCl	250 g
$MgSO_4 \cdot 7H_2O$	10 g
KCl	5 g
$CaCl_2 \cdot 6H_2O$	0.2 g
Yeast extract	10 g
Tryptone	2.5 g
Agar	20 g

The quantities given are to be used in preparation of 1 L final volume of the medium. In preparation make up two solutions, one involving the yeast extract, tryptone, and agar, and the other the salts.

Adjust pH of the agar solution to 7.0. Sterilize the solutions separately at 121° C for 15 minutes. Mix and dispense aseptically approximately 6 ml per 13-mm sterile tube.

Lauryl Sulfate Broth (Double Strength)

Tryptose	40.0 g
Lactose	10.0 g
NaCl	10.0 g
K_2HPO_4	5.5 g
KH_2PO_4	5.5 g
Sodium lauryl sulfate	0.2 g

pH 6.8 (approx.)

Dissolve constituents in 1 liter of distilled water. Dispense 10 ml quantities in large test tubes containing inverted Durham tubes. Sterilize at 121° C for 15 minutes.

Lauryl Sulfate Broth (Single Strength)

Halve the double strength lauryl sulfate broth ingredient amounts.

MacConkey Agar

Peptone	17.0 g
Proteose peptone	3.0 g
Bacto bile salts no. 3	10.0 g
NaCl	5.0 g
Agar	13.5 g
Neutral red	0.03 g
Crystal violet	0.001 g
Distilled water	1000 ml

pH 7.1

This medium is available commercially. Prepare as directed.

Mannitol Salt Agar for Staphylococcus Isolation

Phenol red agar base, reconstituted	1000 ml (Use manufacturer's instructions)
d-mannitol	10.0 g
NaCl	75.0 g

Add the d-mannitol and sodium chloride to the phenol red base while it is still melted, and heat to dissolve. Sterilize at 121° C for 15 minutes.

Methyl Red–Voges-Proskauer Medium

Buffered peptone	7.0 g
Dipotassium phosphate	5.0 g
Glucose	5.0 g
Distilled water	1000 ml

Final pH 6.9 ± 0.2

Dissolve solids in distilled water. Distribute 5 ml/tube and sterilize 15 minutes at 121° C. Two tubes are needed for the test, one for Voges-Proskauer and the other for methyl red.

Mineral Salts + 0.5% Glucose (or Other Sugar)

	1×	10×
$(NH_4)_2SO_4$	1.0 g	10.0 g
K_2HPO_4	7.0 g	70.0 g
KH_2PO_4	3.0 g	30.0 g
$MgSO_4 \cdot 7 H_2O$	0.1 g	1.0 g
Distilled water	1 liter	1 liter

pH 7.0

Sterilize 20 minutes at 121° C.

Glucose agar base:

Agar	6.0 g
Glucose	2.0 g
Distilled water	360 ml

Sterilize for 15 minutes at 118° C.

Add 40 ml of the 10× salts to 360 ml glucose agar before pouring into plates.

Motility Medium

Dehydrated nutrient broth	8.0 g
Distilled water	1000 ml
Triphenyl tetrazolium chloride (TTC)	5 ml of 0.5% aqueous solution
Agar	3.0 g

Mix the dehydrated nutrient broth and agar in the water. Heat until the agar has dissolved and adjust to pH 6.8. Add the diluted dye (see below.) Dispense 10 ml/tube. Sterilize at 121° C for 15 minutes.

Dissolve 25 mg TTC in 5 ml of distilled water. Make up the dye solution just before it is added to the medium.

The final pH of this medium is 6.8 at room temperature. Use screw top tubes to reduce evaporation and extend the storage time of the medium.

Motility medium is also available commercially from Difco Laboratories.

Mueller-Hinton Agar

Beef, infusion from	300.0 g
Casamino acids, tech	17.5 g
Starch	1.5 g
Agar	17.0 g
Distilled water	1000 ml

When using a commercial base:

Mueller-Hinton broth (dehydrated)	21.0 g
Agar	17.0 g
Distilled water	1000 ml

Sterilize 15 minutes at 121° C.

Nutrient Broth/Agar

Beef extract	3.0 g
Peptone	5.0 g
Distilled water	1000 ml
(Agar)	(15.0 g)

pH 6.9 (approx.)

Dissolve the beef extract and peptone in warm distilled water. Add agar for plates, slants, or deeps. Adjust pH if necessary to 6.9. Sterilize at 121° C for 15 minutes. Commercial dehydrated broth is available and very convenient.

Nutrient + Starch Agar

Nutrient broth	1000 ml
Agar	15.0 g
Soluble starch	2.0 g

Prepare nutrient broth. Add agar and heat to melt agar. Suspend the starch in a small quantity of cold water and add to the nutrient agar while it is still hot. pH 7 before and after sterilizing.

Nutrient + Starch + Glucose Agar

Add 20 grams (2%) glucose to above medium.

1% Peptone Agar Deeps

Peptone	10.0 g
Agar	10.0 g
Distilled water	1000 ml

Dissolve and distribute 10 ml/screw cap tubes. Sterilize 15 minutes at 121° C.

Sabouraud's Dextrose Agar

Dextrose	40.0 g
Peptone	10.0 g
Agar	15.0 g
Distilled water	1000 ml

Final pH 5.6 (approx.)

For Sabouraud's dextrose broth omit the agar.

Saline

NaCl	8.5 g
Distilled water	1000 ml

Simmons Citrate Agar

Magnesium sulfate	0.2 g
Monoammonium phosphate	1.0 g
Dipotassium phosphate	1.0 g
Sodium citrate	2.0 g
NaCl	5.0 g
Agar	15.0 g
Brom thymol blue	0.08 g

To rehydrate the medium, suspend 24.2 g of Simmons citrate agar in 1000 ml of cold distilled water and heat to boiling to dissolve the medium completely. Distribute in tubes or flasks and sterilize by autoclaving for 15 minutes at 121° C. Allow the tubed medium to cool in a slanting position. The pH of the final reaction of the medium will be 6.8

Sugar Fermentation Tubes of Bacteria

Nutrient broth (dehydrated)	8.0 g
Glucose or other sugar	5.0 g
Andrade's indicator*	10 ml
Water	1000 ml

pH 7.1 before sterilizing
pH 7.0 after sterilizing

Add 6 ml/standard tube. Add Durham tubes inverted before sterilizing—they will fill in the autoclave. Sterilize for 12–13 minutes at 118° C. The procedure for fermentation tubes is listed below.

*Andrade's indicator:

Acid fuchsin   1.0 g

Distilled water 750 ml

1 M NaOH

To dissolve and neutralize the acid fuchsin solution, add 7 ml 1M NaOH. Allow to stand for several hours. If the solution is still red, add 3–4 ml more of 1M NaOH. Let stand for several hours. If solution is still red, add 1–2 ml more. Continue this until solution is light red. Let stand for several hours and adjust pH to about 8.0—it will still appear a light red. (When added to medium, the broth will be a very pale yellow.)

This can be stored for long periods at room temperature. Although the preparation takes time, it is a good indicator for fermentation because it changes to red at about pH 6.8. There are many formulas for this indicator, but this has worked well in our media.

Alternative Fermentation Tubes: Phenol Red Broth Base for Carbohydrate Fermentations

Beef extract	1.0 g
Proteose peptone no. 3	10.0 g
NaCl	5.0 g
Phenol red	0.018 g

Final pH 7.4

Dissolve 16 g phenol red broth base in 100 ml of distilled water. Add the proper amount of carbohydrate desired (usually 0.5% or 5 g per liter), distribute in fermentation tubes with small inverted Durham tubes. Fill these tubes with fluid by autoclaving in a sterilizer for 15 minutes at

121° C. It is recommended that you test the first batch of carbohydrates with known fermenters. Occasionally the carbohydrate hydrolyzes. If this occurs, lower temperatures and shorter sterilization time are recommended, testing for sterility.

Thioglycollate Medium, Fluid

Yeast extract	5.0 g
Casitone	15.0 g
Dextrose	5.0 g
NaCl	2.5 g
1-Cystine	0.75 g
Thioglycollic acid	0.3 ml
Agar	0.75 g
Resazurin, certified	0.001 g

Final pH 7.1

Suspend 29.5 g of the thioglycollate medium in 1000 ml cold distilled water, and heat to boiling to dissolve completely. Distribute into tubes and sterilize in the autoclave for 15 minutes at 121° C. This medium can be obtained from Difco or BBL.

Tryptone Broth

Tryptone	10.0 g
NaCl	5.0 g
Distilled water	1000 ml

pH 7.0 before sterilizing

Mix and sterilize for 15 minutes at 121° C.

Tryptone Agar for Base Plates

Tryptone	10.0 g
NaCl	5.0 g
$MgSO_4 \cdot 7H_2O$	1.0 g
Agar	15.0 g
Distilled water	1000 ml

pH 7 before sterilizing

Autoclave 121° C for 15 minutes.

Tryptone Soft Agar Overlay

Same formula as above but use 7 grams of agar instead of 15 grams.

Trypticase (Tryptic) Soy Broth (TSB)

Trypticase peptone	17.0 g
Phytone peptone	3.0 g
NaCl	5.0 g
K_2HPO_4	2.5 g
Glucose	2.5 g
Distilled water	1000 ml

pH 7.3 (approx.)
Sterilize at 121° C for 15 minutes.

Trypticase (Tryptic) Soy Agar (TSA)

Trypticase soy broth: as above or 30 g of
commercial dehydrated mix
Agar   15.0 g
Distilled water   1000 ml (needed for rehydration
                                of commercial mix)

Sterilize at 121° C for 15 minutes.

For TS + streptomycin plates: Add 1.0 ml of
stock streptomycin per liter of sterile, melted,
cooled agar (stock streptomycin: 1.0 g
streptomycin/10 ml distilled water, filter
sterilized).

Trypticase Soy Yeast (TSY) Extract Agar

Trypticase soy broth	30.0 g
Yeast extract	10.0 g
Distilled water	1000 ml
(Agar)	(15.0 g)

pH 7.2 before sterilizing

Add agar for plates, slants, or deeps.

TSY + Glucose Agar

Add 2.5 grams glucose/liter.

TSY + Glucose + Brom Cresol Purple Agar Slants

Trypticase soy broth	30.0 g
Yeast extract	10.0 g
Glucose	17.5 g
Brom cresol purple	0.04 g
Agar	15.0 g
Distilled water	1000 ml

pH 7.3 before sterilizing
pH 7.0 after sterilizing

Sterilize at 118° C for 15 minutes, 6 ml/tube.

TYEG Salts Agar Plates Containing 0.5, 5, 10, and 20% NaCl

For basal medium composition see TYEG salts agar.
The two lower salt concentrations (0.5 and 5%)
can be added directly to the basal medium, whereas
the two higher salt concentrations (10 and 20%)
need to be autoclaved separately and added back to
the basal medium after autoclaving. For related
procedural details see the ATCC agar medium 217
procedure.

TYEG Salts Agar Plates Containing 0, 10, 25, and 50% Sucrose

For basal medium composition see TYEG salts agar.
The three 10% sucrose concentration can be added
directly to the basal medium, whereas the
remaining two sucrose concentrations (25 and
50%) need to be autoclaved separately and added
back to the basal medium after autoclaving. For
related procedural details see the ATCC agar
medium 213 procedure.

TYEG (Tryptone Yeast Extract Glucose) Salts Agar Slants

Tryptone	10 g
Yeast extract	5 g
Glucose	2 g
K_2HPO_4	3 g
$CaCl_2$-$6H_2O$	0.2 g
Agar	20 g
Distilled water	1000 ml

pH 7.0 (before autoclaving). Dispense 6 ml/tube. Autoclave 121° C for 15 minutes.

Urea Agar

Urea agar base	29.0 g

Ingredients in this base:

Peptone	1.0 g
Dextrose	1.0 g
NaCl	5.0 g
Monopotassium Phosphate	2.0 g
Urea	20.0 g
Phenol red	0.012 g
Agar	15.0 g

Final pH 6.8

Suspend 29 g of urea agar base in 100 ml of distilled water. Filter-sterilize this concentrated base. Dissolve 15 g of agar in 900-ml distilled water by boiling, and sterilize for 15 minutes at 121° C. Cool to 50° to 55° C and add 100 ml of the filter-sterilized urea agar base under aseptic conditions. Mix and distribute in sterile tubes. Slant tubes so they will have a butt of 1 inch and slant 1½ inches in length, and solidify in this position.

Worfel-Ferguson Slants

NaCl	2.0 g
K_2SO_4	1.0 g
$MgSO_4 \bullet 7H_2O$	0.25 g
Sucrose	20.0 g
Yeast extract	2.0 g
Agar	15.0 g
Distilled water	1000 ml

No pH adjustment

Sterilize 15 minutes at 121° C.

Yeast Fermentation Broth Containing Durham Tubes

Tryptone	10.0 g
Yeast extract	5.0 g
K_2HPO_4	5.0 g
Sugar (glucose, lactose, or maltose)	10.0 g
Distilled water	1000 ml

The method of preparation is the same for phenol red broth base. *Note:* Be careful not to overautoclave, because the sugars may decompose with overheating—121° C for 15 minutes.

APPENDIX 4

Major Capital Equipment Needs, Special Glassware, and Microscopic Prepared Slides

Exercise 2

Light microscope, monocular or binocular, with low, high, and oil-immersion objectives, an Abbe-type condenser, and a suitable light source

Exercise 3

Star diaphragm for dark-field microscopy
Hanging drop depression slides

Exercise 9

GasPak system

Exercise 10

Spectrophotometer, colorimeter or Klett with adapters to fit standard tubes
Two water baths

Exercise 12

UV lamp with shielding
Safety glasses for use with UV lamp
Dark box for storing plates

Exercise 16

Water bath

Exercise 20

Dissecting microscope

Exercise 22

Membrane filter apparatus

Exercise 23

GasPak anaerobe jar
Magnifying glasses or dissecting microscope (optional)

Exercise 24

Candle jar

Exercise 27

Coplin jar with 95% denatured ethanol

Exercise 29

Gooseneck, fluorescent, or fiber optic lamp
Hypodermic needle without bevel, 18 gauge
Syringe, Luer-type, 1–2 ml
Serology test tubes

Exercise 30

Vortex mixer
High-intensity incandescent light source
Mechanical rotator for slides (optional)

Exercise 31

GasPak jars.

Exercise 32

Commercial light box providing indirect light against a dark background
Reagent quality water

Exercise 34

Membrane filter apparatus
Vacuum pump or Venturi vacuum system

Equipment per Student

Laboratory coat
Waterproof pen
Loop

Equipment per Team (preferably 2)

Bunsen burner with pilot light
Tripod and/or ringstand
Beaker or can
Test tube rack
Glass slides
Glass coverslips #2
Bulbs and/or propipets
Snap clothespin
Pipets, 1.0 ml and 10.0 ml
Small widemouth jars or small dilution bottles
  filled ½ with fine sand and covered with 95%
  alcohol
Forceps

Available in the Laboratory

Plastic bins and autoclave bags for discard
Staining bars to hold slides over sinks and beakers
  or cans
Bent glass rods: a 15-cm (6-inch) glass rod bent in
  the shape of a hockey stick
Jars of 95% ethanol for sterilizing bent glass rods
  and for other uses
Vaseline
Concave hanging drop slides
Discard buckets for pipets
Laboratory wipes such as Kimwipes
Sterile cotton swabs
Lens paper
Immersion oil
Bottles with stains (250-ml size convenient)
Mild cleanser such as Boraxo or Bon Ami
Dye remover for removing stains from hands
Ice bucket
Thermometers
Wash bottles
Toothpicks/applicator sticks
Rulers (divided in mm)
Scalpels and/or sharp knives
Mortar and pestle
Cotton balls
Tongue depressors

Laboratory

Hazardous waste container
Fire extinguisher
Safety eyewash
Disinfectant for wiping down surfaces
Hand washing soap
Paper towels
Incubators, 37° C (or 35°) and 30° C (if convenient)
Water baths

Media Preparation

Autoclave
Source of distilled water and sterile water
pH meter
  pH 7 standard (and other standards)
  1 M HCl
  1 M NaOH
Hot plate
Balance accurate to 10 mg
Apparatus for dispensing media such as a Cornwall
  syringe-type pipetter
Microwave oven for melting agar
Magnetic stirrer (optional)
Refrigerator for storing cultures and reagents
Freezer (without a self-defrost cycle) for storing
  cultures
Water bath for cooling agar before pouring plates
Membrane filter apparatus (disposable filters for use
  with a syringe are very convenient for small
  amounts)
Filter paper discs
Large containers for mixing media
Large spoons, spatulas, and stirring rods
Graduated metal pitchers for dispensing media
  (optional, but convenient)
Erlenmeyer flasks
Beakers
Standard (15-cm) and small test tubes; some with
  plastic caps for class and some with screw caps for
  storing cultures
Petri plates
Durham tubes

Microscopic Prepared Slides

The following slides may be obtained from Carolina Biological Supply Company.

Exercise 2

Ba 234 *Staphylococcus*
Ba 050 *Bacillus:* typical rod forms
Ba 030 *Spirillum:* comma, spiral forms

Exercise 5

Bacterial flagellar stains (*Alcaligenes faecalis* and *Pseudomonas fluorescens*)

Exercise 21

18 slides of different protozoans and worms
Subkingdom Protozoa
Phylum Sarcodina (pseudopodia)
  1. *Entamoeba histolytica* trophozoite
  2. *Entamoeba histolytica* cyst
Phylum Ciliata (cilia)
  3. *Paramecium* trophozoite
Phylum Mastigophora (flagella)
  4. *Giardia lambia* trophozoite
  5. *Giardia lambia* cyst
  6. *Trypanosoma gambiense*

Phylum Sporozoa (nonmotile)
  7. *Plasmodium vivax* ring
  8. *Plasmodium vivax* amoeboid
  9. *Plasmodium vivax* schizont
Subkingdom Helminths (worms)
Phylum Platyhelminthes (flatworms)
Class Turbellaria (free living)
  10. *Dugesia* sp.
Class Tremetoda (flukes)
  11. *Schistosoma mansoni* adult male
  12. adult female
  13. ovum (egg)
  14. infective ciliate cercaria
  15. ciliated miracidium
  16. sporocyst stage in snail liver tissue
  17. *Clonorchis sinensis*
Class Cestoda (tape worms)
  18. *Taenia solium* trophozoite

Exercise 27

H1150 Human blood smear (unstained)
H1155 Human blood smear (stained with Wright's stain)
H6445 Bone marrow (stained with Wright-Giemsa; contains a greater variety of WBC)

APPENDIX 5

Notes to the Instructor

Preparation for Labs

Frequently it is necessary to inoculate a large number of tubes. An efficient method of doing this is to add about 5–10 ml of sterile water to an overnight slant of bacteria. Suspend the cells with a pipet. Then inoculate each slant with a few drops of the suspension by wiggling a pipet up the surface of the slant.

In the lab manual, glass beakers are used to melt agar or steam slides. Metal cans, however, are really superior because they don't break when heat is applied directly to them, they are very inexpensive, and they don't have to be washed.

A loop is a very important tool. They can be made from 24-gauge nichrome wire and a soft aluminum tubing of about 5 mm in diameter. Cut the tube for the handle about 15 cm (6 inches) long and the wire about 13 cm (5 inches) long. Crimp the wire in the tube, smooth other end, and form a loop in the wire of about 4 mm (⅛ inch) in diameter. The wire should be long enough to inoculate a slant without contaminating the aluminum handle.

Exercise 1

1. If it is more convenient, incubate organisms from the body at room temperature too. They will not grow as well, but will be fine for this exercise after seven days. One of the reasons to incubate at 37° C is to illustrate the principle that not all organisms grow best at the same temperature, but usually grow optimally at the temperature of their natural habitat.
2. It is helpful to have demonstration cultures of bacteria and molds.

Exercise 2

1. Since the "Getting Started" section of this exercise is rather long you may wish to assign it as outside reading emphasizing the following areas:
   a. Understanding of definitions.
   b. Understanding working principles of bright-field microscopy.
   c. Familiarity with precautions for proper microscope care.
2. When introducing students to the use of the oil-immersion objective, you may rather substitute a prepared stained slide for a wet mount. A beginner's problem with a wet mount preparation can be using too large a drop of the culture suspension. This often causes the coverslip to lift off the slide surface when focusing upward with the oil-immersion objective.
3. A working replica of van Leeuwenhoek's microscope may be obtained by writing to the Dept. of Microbiology, Montana State University, Bozeman, MT, 59717. A variety of brass replicas, ranging in cost from $40 to $80, are available.

Exercise 3

1. For precautions with wet mounts involving the use of the oil-immersion see objective exercise 2, "Notes to the Instructor."
2. If time is a factor you may wish to set up the dark-field microscopy section as a demonstration. In the event a star diaphragm is not available a series of opaque circular cardboard discs can readily be prepared for insertion in the filter holder. If a phase microscope is available the dark-field effect

may also be viewed with the low power objective of the phase microscope simply by rotating the phase ring to the proper position.

Exercise 4

1. Stains of smears made from cultures growing on agar slants are much easier to see than ones made from broth cultures. It is better to begin staining with cultures from agar slants and introduce staining broth cultures after students are more experienced.
2. Students can use a clothespin (snap-type) to hold their slide over the sink as they are staining it. When staining more than one it may be better to place them on staining bars made from glass tubes connected with plastic tubing. It is easy to rinse the slides over the sink with water if a small hose is attached to the faucet.
3. Since it is almost inevitable that students' hands will be stained, it is helpful to have a dye remover available for their use. Students can wear latex gloves, but they are a bit awkward to work with.
4. The negative stain can be considered optional. The method presented here is a simplified version. See *Manual for General and Molecular Bacteriology*, Philip Gerhardt, editor in chief, for a better preparation but more complicated procedure.
5. Common problems of students first observing material under the microscope:
   a. Slide is in focus on high-dry, but can't be put in focus using oil-immersion lens. *Solution:* Check to see if slide is upside down. Also, some oil-immersion objective lenses are spring-loaded and get stuck in the up position.
   b. Bubbles appear to come up through the oil. *Solution:* Oil was added before stain was dry. It may be necessary to start over.
   c. Bacteria appear distorted and indistinct under the oil-immersion lens. *Solution:* There may not be enough oil on the slide—add more oil. Distortion can also be caused by heat-fixing the slide before the smear is dry; it will be necessary to start over.

Exercise 5

1. Cans for boiling the water instead of beakers are particularly useful for this exercise. Metal bars or U-shaped glass rods can be placed over the can to hold the slides.

2. The cultures for the Gram stain are predominantly gram positive because these cultures happen to have more interesting arrangements of cells. More gram-negative cultures can be added or substituted.
3. The acid-fast stain is optional here. It would probably be more interesting in a later class when tuberculosis and leprosy are discussed.
4. The Schaeffer-Fulton method for staining endospores uses malachite green and is much preferable to carbol fuchsin. The water is not supposed to boil, but boiling is easier for students and does not seem to affect the results. Dye should be continually added to prevent it from drying on the slide.
5. *Bacillus cereus* (or other *Bacillus*) can be encouraged to sporulate by growing on nutrient agar for 1–4 days. Most *Bacillus* sporulate poorly on TS agar because it is so rich. Try to give the cultures to the students when the endospores are still within the cells.
6. To prepare a culture for an acid-fast stain, grow *Mycobacterium smegmatis* on TS agar and then make a heavy suspension of cells in saline. Add suspensions of other bacteria from slants so that there will be a mixture of acid-fast and non-acid-fast organisms. Add a few drops of phenol to the mixture and store it in the refrigerator. It will keep for years.
7. *Spirillum* grows well in nutrient broth, but it is easy to maintain in stabs of cytophaga agar.
8. Worfel-Ferguson Agar can be used for the detection of capsule production by *Klebsiella*.

Exercise 6

1. Grow separate broth cultures of *Staphylococcus epidermidis* and *Micrococcus luteus* for 24 hours and incubate at 37° C. Thoroughly mix together about three parts of *Micrococcus* to two parts *Staphylococcus*. Dispense about 1 ml/small tube. Students should be sure to shake tubes gently before streaking because the *Micrococcus* have a tendency to settle to the bottom.
2. Loops sometimes become rough from repeated heating in a flame. Smooth them by rubbing with a piece of sandpaper. Also, students should check to be sure there isn't a gap in the loop that might dig into the agar.

Exercise 7

1. *E. coli* produces the unique green sheen on EMB agar because it produces a large amount of acid during fermentation. The acid complexes with the dyes in the agar.

Exercise 8

1. Inoculate *E. coli* into TS broth and incubate at 37° C overnight without shaking. Before class dilute 1:1 with TS broth and label Suspension A. Make a 1/100 dilution of Suspension A and label it Suspension B, which should appear clear. Pipette a few ml of each suspension into a labeled tube for each team.

2. Since an overnight culture of *E. coli* K12 has about 1×10^9 cells/ml, Suspension A should have about 5×10^8 bacteria/ml and Suspension B should have about 5×10^6 bacteria/ml. If it is convenient for you to use another strain of *E. coli* that grows to a higher density overnight, simply dilute Suspension B until it is clear.

3. Refrigerate cultures before class or they will continue to grow.

4. It might be helpful if you again emphasize how many organisms can be in a liquid that appears clear.

5. Errors include pipetting errors, errors in volume of water blanks, counting errors, etc. It is much more important to accurately measure a small volume than the larger volume of the blank.

6. Serial dilutions of $1/10 \times 1/10 \times 1/10 = 1/1000$ or 10^{-3}.

7. Adding 1 ml to 99 ml = 1/1 + 99 = 1/100 or 10^{-2}.

8. No matter how a solution is diluted, simply invert and multiply the numbers of colonies on the plate and you will have the original concentration/ml. For example, if a sample is diluted 1/32, which results in 64 colonies from 1 ml, multiply 32×64 to obtain original concentration of 2048 organisms/ml.

Exercise 9

1. Inoculate one set of TS broth (5 ml/tube) with *E. coli* (label A, for example) and another set with *Micrococcus* (label B). Incubate 24 hours at 37° C. Before inoculating the *Clostridium*, boil the tubes of TS broth five to ten minutes to remove the air. Let cool, inoculate each tube with a drop of stock *Clostridium* growing in thioglycollate broth or BHI (brain-heart infusion) and incubate in a GasPak jar at 37° C for 48 hours (label C). If a GasPak jar is not available for the *Clostridium*, inoculate thioglycollate broth or BHI broth (12 ml/tube) after heating to remove air and incubate for 24–48 hours.

2. After the class has inoculated the deeps, incubate at room temperature. At 37° C *E. Coli* sometimes produces so much gas that the agar is blown apart. More than two-day incubation may be necessary for good growth—ours are incubated five days.

3. Glucose is included in the media to encourage anaerobic growth, but it might not be needed for some strains, and less gas would be produced without it.

4. The results in this exercise may not agree with the diagram in the Nester text (fig. 4.9). The text probably indicates the appearance of colonies if the agar tubes were not boiled first.

5. The GasPak system for obtaining anaerobic conditions is expensive but very useful. In addition to the jar, you must purchase envelopes, pH indicator strips, and catalyst. The catalyst may be reused by baking in an oven for two hours at 160° C to drive out moisture. Alternately, the catalyst can be stored in a 55° C (130° F) incubator after use for at least 24 hours.

Exercise 10

1. This is an excellent exercise to teach students how to gather data and plot the results. In this exercise the effects of two different temperatures on generation time are compared, but other variables such as rich and minimal medium or different strains of bacteria could be compared as individual or special projects.

2. The plot of the O.D. of organisms growing in log phase will be straight line on semi-log paper and a curved line on conventional graph paper. If the log is taken of the O.D. readings and then plotted on conventional graph paper, a straight line will also be generated. You can use this method if you prefer. Log paper is just a very convenient method for plotting growth curves.

3. Since the data are supposed to generate a straight line on semi-log paper, it is appropriate to draw a "best-fit" line with a straight edge, thereby averaging the data.

4. A spectrophotometer and Klett-Sommerson colorimeter are equally useful for this exercise. Be sure it has an adapter to fit a standard test tube; it is very awkward to transfer a portion of the growing culture to a special tube or cuvette for readings. Adapters are made to fit several different-sized tubes for the Klett and most spectrophotometers (contact the company or check the catalog). A Klett reading is O.D. (or absorbance) × 500 because evidently the inventor thought large numbers were easier to use.

5. When labeling semi-log paper, the 1 on the bottom line can be 0.01, 0.1, 1.0, etc., and the 1 line on the next cycle up would be 0.1, 1.0, 10 respectively, depending on the range of numbers graphed. Remind the students that the value of the spaces changes as the numbers on the Y axis become larger.

6. To obtain cells in log phase, dilute an overnight broth culture 1:5 into TS broth about 1-½ hours before class and incubate at 37° C.

7. Prewarm the tubes the students will use to measure the growth rate so they will not have to equilibrate during the initial readings.

8. It is important that the cultures are out of the water bath as little as possible, so avoid long lines of students waiting to use the spectrophotometers. The time between readings isn't important as long as the exact time is recorded.

To calculate the generation time:
   a. Find the convenient interval in which the optical density doubles: for example, from 1 to 2, or on the example graph, from .08 to .16.
   b. Determine on the horizontal X axis how many minutes it took the culture to double.
   c. The generation times for the cultures in the example graph are 40 minutes at 30° C and 30 minutes at 37° C.

Exercise 11

You may wish to substitute another spore-forming bacillus for *B. subtilis*—for example, *B. cereus* (see ex. 5) or *B. megaterium*. All of these are mesophiles and grow rapidly on complex media such as nutrient agar and TS agar. *B. megaterium* offers the advantage of a large cell which is readily seen under the microscope.

About two weeks before initiation of the exercise, the instructor should inoculate a nutrient agar slant with a spore-forming bacillus and incubate the slant at 30° C for 1–5 days, then examine the culture for the presence of endospores.

Exercise 12

To prepare a suspension of *Bacillus cereus* spores in sterile saline, inoculate a nutrient slant with *B. cereus*, incubate 4–5 days at 30° C, suspend the cells in approximately 10 ml of sterile saline, and transfer the cell suspension into sterile 5-ml dropper bottles. The bottles can be refrigerated until ready for use in the classroom.

Sylvania produces a relatively inexpensive 30-watt low-power (118 v., 0.36 amp) UV lamp which is 21 inches long and 9 inches high. Ask for Sylvania engineering bulletin G-342 for details about other models of their UV lamps. For a supplier, contact your local hardware store.

Safety glasses are necessary only if you have a UV lamp which is not properly shielded.

Exercise 13

1. Sterilize the media containing sugar for only 10 minutes at 121° C rather than the traditional 15 minutes for most media. This is because sugars hydrolyze to some extent during autoclaving. For sterility control purposes a tube of the autoclaved medium should be incubated at 37° C for 48 hours.

2. Depending on available time you may wish to reduce or, if necessary, eliminate the wet mount study. You could ask each of a pair of students to prepare wet mounts of only two of the four test organisms. For additional reduction you could ask all students to prepare wet mounts of only one organism—possibly the one showing growth on the highest sugar and salt concentration.

Exercise 14

1. The *S. epidermidis* and *E. coli* broth cultures may be used for both the antiseptic and the antibiotic studies.

2. Only two antiseptic agar plates need to be prepared per group of 2–4 students in that each student will have an opportunity to prepare at least one antibiotic filter paper disc agar diffusion plate.

3. Hexachlorophene (brown granules) are available from Sigma Chemical Company. It dissolves readily in ethanol. For a 0.3% solution dissolve 0.3 g in 100 ml ethanol.

4. If more than 4 antiseptic discs are placed on an agar plate they should be arranged in a manner similar to that used with the antibiotic discs (see fig. 14.3). In such an instance it might be better to use ¼-inch rather than ½-inch filter paper discs.

5. In order to save time, the antibiotic disc assay procedure could be reduced to as little as 2 to 3 microorganisms, e.g., *S. epidermidis* (gram positive), *E. coli* (gram negative), and *Pseudomonas aeruginosa* (gram negative, high resistance). You may wish to include a fungus, e.g., *S. cerevisiae*, to demonstrate resistance to antibiotics other than nystatin.

Exercise 15

1. Be sure the strain of *E. coli* you use for this exercise is sensitive to streptomycin as strains of *E. coli* vary. *E. coli* K12 has worked well.

2. Incubate the broth culture used for the class for 24 hours. The higher the population, the more mutants will be seen.

3. Streptomycin can be diluted 1.0 gram/10 ml distilled water for a concentration of 100 mg/ml, the same as for exercise 16. Filter-sterilize and freeze in convenient amounts. For class, dilute the stock 1/3.3 to make a concentration of 30 mg/ml (1 ml into 2.3 ml sterile distilled water). The final concentration in the agar or broth is about 250 µg/ml. Streptomycin is fairly stable, but store in an ice bucket during the class. It is convenient to prepare enough for both labs and store in refrigerator. To encourage students not to waste the streptomycin, it can be worthwhile in the first session to dispense about 0.6 ml/small tube, which is enough for one team.

4. Incubate the bottles or flasks at least 5 days because it usually takes longer for the streptomycin-resistant mutants to grow. Be careful that the plates don't dry up during the long incubation.

5. In the second session, it is efficient for the students to come to the instructor with their melted, cooled agar. The instructor or the student can add 0.1 ml streptomycin to the tube.

6. Sometimes a swirl of little colonies will be seen on the edge of the plate with streptomycin. These are probably the result of insufficient mixing.

7. Strains of *E. coli* K12 may vary in their sensitivity. If no mutants can be seen with your strain, try a concentration of 100 µg streptomycin/ml.

Exercise 16

1. This is a very interesting, inexpensive, and easy exercise to demonstrate transformation. It is important that the instructor be unusually enthusiastic about what is really happening—the students will be changing the actual genes of one organism by immersing it in the DNA of another. Unfortunately, all that can be seen on the plates are patches of growth or no growth, but the significance of these results is very important.

2. This exercise is an excellent example of the use of controls. Some additional ones some students might like to run follow:
   a. Is the DNase preparation sterile? (Add DNase alone to one section of agar)
   b. Does DNase kill intact cells? (Mix DNase and cells on one section of agar)

3. The standard saline citrate is at 10 × concentration so that students can add 1 ml of culture to a 0.1 ml solution to bring it to the correct concentration.

4. Incubate these plates at least 48 hours at 37° C. It seems to take the transformed cells longer to grow than the streptomycin-resistant strains on agar containing streptomycin.

5. Prepare a stock solution by adding 1.0 gram streptomycin to 10 ml distilled water and filter-sterilize through a membrane filter of 0.45 micrometers (this can be frozen). Add 1.0 ml of the stock solution to a liter of autoclaved, cooled TSY agar before pouring plates. The final concentration will be 100 µg/ml. Store the plates at room temperature for a day or until the surface is dry, then store in refrigerator.

Exercise 17

1. Cultures should be in log phase for best results. Dilute an overnight culture about 1/5 two hours before class and incubate at 37° C. Do not shake cultures because the pili are fragile.
2. The bent glass rod can either be sterilized in an autoclave bag, one rod to a bag, or the students can dip it in alcohol and then burn off the alcohol by touching it to a flame. Warn them not to let the burning alcohol touch their hand.
3. To calculate the % recombinants, divide the number of recombinants by 1×10^8.
4. You would expect only 1 back mutant/ml if the back mutation was 1×10^{-8} ($1/10^8$).
5. If either culture A or B can grow on MM agar (therefore are not auxotrophs) there is no way to determine how many colonies are the result of conjugation.

Exercise 18

1. It is convenient to put the broth culture of *E. coli* in sterile glass dropper bottles so students can add a few drops to each tube. Be sure the dropper bottles can be autoclaved. Cultures of *Enterobacter* and *Klebsiella* might also be tested.
2. Some strains of *Bacillus* spread over the plate, making it difficult to read the results. This can be avoided somewhat by using dry plates and making a small point inoculation.
3. The iodine used in the Gram stain can be used for this test.
4. *Bacillus* strains vary in their ability to hydrolyze starch and demonstrate catabolite repression, so it might be interesting to try several. Usually *B. cereus* is starch positive and *B. sphaericus* is starch negative.

Exercise 19

If a 25° C incubator is not available the cultures may be incubated at either room temperature or 30° C.

Saccharomyces cerevisiae generally produces asci containing four ascospores when cultivated on glucose-acetate agar. However, the number of ascospores formed per ascus can vary depending on the yeast strain evaluated and composition of the sporulation medium. Another sporulation medium that has enjoyed varying degrees of success is a sterile carrot wedge. You may wish to inoculate a carrot wedge for purposes of comparison with acetate agar. If so, inoculate a test tube containing a sterile carrot wedge with several loopfuls of the *Saccharomyces cerevisiae* inoculum. Incubate at 25° C, and examine carefully for asci and ascospores at the end of one week.

Make certain the inocula are well suspended before making transfers. The germ tube study should be prepared at the beginning of the class period since it must be observed after two hours of incubation.

Exercise 20

1. Try to prepare the Petri plate cultures approximately 5 to 7 days before the classroom period because both *Aspergillus* and *Penicillium* are slow-growing fungi. The genus *Rhizopus* grows somewhat faster and may be inoculated 2 to 3 days prior to the classroom period.
2. Covered slide cultures of the three fungi (*Rhizopus*, *Aspergillus*, and *Penicillium*) should be prepared 1 to 2 days before the classroom period. If time permits, you may wish to have the students prepare their own covered slide cultures. One method for producing covered slide cultures is the following:
   a. Using sterile tweezers (prepared by dipping the tip in alcohol and flaming), aseptically transfer a piece of sterile filter paper to the bottom of a Petri dish. The paper should be approximately the same diameter as the bottom of the plate. If glass plates are used, the paper may be placed in the plate and sterilized in the oven, but care should be taken not to char the paper.
   b. Clean a glass microscope slide, and sterilize it by immersing it in alcohol with a pair of tweezers and then burning off the alcohol. Place the slide on the surface of a laboratory tabletop that was previously wiped with a disinfectant solution.
   c. Prepare some capillary glass tubing, approximately 1 to 2 mm in diameter, and break the tubing into lengths approximating the width of the glass slide. A Pasteur pipet is often a convenient source of tubing. The technique for preparing the capillary tubing simply involves heating the tubing with the flame of a Bunsen burner until the glass begins to soften and then pulling the tubing in both directions.
   Note: Slowly rotate the tubing while heating.

d. Using tweezers, place two pieces of the capillary tubing in a perpendicular manner on a central surface area of the glass slide such that the width separating the two pieces approximates the width of a glass coverslip. Each piece of capillary tubing should be sterilized by flaming in alcohol before being placed on the slide.

e. Sterilize a clean coverslip using an alcohol flaming technique, and transfer it aseptically to the surface of the two pieces of capillary tubing so that the coverslip is supported on either side by capillary tubing.

f. Using an artist's paintbrush and melted vaspar (to melt, place a tube of vaspar in a hot water bath), ring three sides of the coverslip with vaspar, leaving the fourth side open to the atmosphere. Vaspar is prepared by adding equal parts of Vaseline and paraffin to a beaker which is then placed in a hot water bath. After melting, transfer 10- to 15-ml amounts to test tubes for dispersal to students.

g. Melt a tube of agar medium, and cool it to 45° C in a water bath. With an inoculating loop, inoculate the tube of melted agar with a spore suspension of the test culture.

h. Place the tip of a sterile Pasteur pipet into the tube of inoculated agar, and remove a small amount (approximately 0.5 to 1.0 ml) by capillary action.

i. Touch the tip of the Pasteur pipet to the open edge of the coverslip-slide preparation so that the agar flows beneath the coverslip surface by capillary action.

j. Place the slide culture on the surface of the filter paper in the Petri dish.

k. Moisten the filter paper with 2 to 4 ml of sterile water to increase the relative humidity and prevent the agar medium from drying out.

l. Incubate at 25° C (or room temperature). If possible, observe daily, using the low power objective of the microscope. Moisten the filter paper with additional sterile water when necessary. It is often possible to detect germination of the asexual spore after 6 to 12 hours.

Exercise 21

Students should be able to complete the observation part of this exercise in a 3-hour laboratory period. Alternatively, if there is less than a 3-hour laboratory period, either the laboratory could be extended into another period or some parts of the exercise could be omitted. Pursuit of the latter alternative might prove of less value to the students.

Exercise 22

1. The instructor can assign one team to make a control plate of *E. coli* without adding any phage for a negative control—it seems unnecessary for every team to make a control plate. If a high titer of a phage known to infect *E. coli* K12 is available, a drop might be placed on the lawn for a positive control. *Note:* A good special project would be the preparation of a high-titer phage suspension for use as a control next year (see number 5 below).

2. It is convenient to put the bacteria in glass dropper bottles so that the student can simply add a few drops instead of using a pipet.

3. Use great caution when filtering sewage—it can contain pathogenic bacteria and viruses. Wear disposable gloves and work in a hood. If it isn't possible to filter the sewage, let the sample settle, and remove material from the top. Add chloroform and mix gently. The chloroform will kill the bacteria (including *Bdellovibrio*), but most phage are resistant.

4. *Bdellovibrio* doesn't live long without a host, so the sewage should be filtered just before class to increase the chances of isolating these organisms.

5. Preparing a phage stock would be an interesting project for a student. Phage can be transferred from a plaque to a few ml of broth by stabbing a loop or straight wire into the center of the plaque and then agitating it in the broth. The whole plaque can also be transferred with the wrong end of a pipet. Add a few drops of chloroform to kill the bacteria and store in the refrigerator (do not freeze) until needed.

6. Another interesting project would be isolating phage plaques using a streak-plate technique. Stab a loop in a plaque and then streak onto a

tryptone hard base plate (as you would for isolating bacteria). Add 0.1 ml of an overnight *E. coli* K12 to 4 ml of melted and cooled overlay agar and pour over plate. Incubate at 37° C and examine for isolated phage plaques.

Exercise 23

1. Contact plates are small Petri plates that are filled with agar to the brim. This takes a little practice. They can be pressed directly on the skin, the cover replaced, and incubated.
2. Some gram-positive cocci are microaerophilic and will be seen growing in a band a few cm below the surface of the agar.
3. Check anaerobe jars about 30 minutes after closing. If a condensate has formed on the sides of the jar, the jar will probably be anaerobic. It takes a few hours for the jar to become anaerobic and the indicator to change to white.
4. Propionibacteria must be incubated anaerobically at least 5 days.

Exercise 24

1. Demonstration cultures of α-hemolytic *Streptococcus* and β-hemolytic *Streptococcus* on blood agar plates are very useful. Incubating them in a candle jar increases the CO_2 and usually gives better hemolysis, but is not necessary. If desired, big institution-sized pickle jars make good candle jars. Simply add inverted plates and light a candle in the jar. Screw the lid on tightly and the candle will soon go out.
2. Sheep blood is preferable for blood agar plates—see appendix 3.
3. α- and β-hemolytic streptococci can be purchased. They have to be transferred every four weeks, so it is convenient to freeze them.
4. Emphasize that beta hemolysis can be produced by several organisms and does not always indicate a pathogenic organism. Here we are only interested in beta hemolysis caused by *Streptococcus pyogenes*.

Exercise 25

1. It is useful for the students to generate the class Gram-negative identification chart from the results of their "known" organisms. Put a chart on the blackboard for the students to record the results of their known organism, which can then be used to identify their unknown. This gets around the problem of some bacterial strains having variable fermentation patterns, motility, or other noncritical characteristics. For instance, some *E. coli* strains ferment sucrose, and others don't.
2. If students don't agree on the results, it may be necessary for them to repeat the tests.
3. The tests indol, methyl red, Voges-Proskauer, and citrate are collectively called the IMViC test and have been traditionally used for identifying coliform organisms.
4. It is very helpful to have cultures demonstrating positive and negative reactions.
5. The Voges-Proskauer test sometimes takes 30 minutes to turn positive.

Exercise 26

1. This can be a very interesting exercise for the students and instructor. When deciding on mixtures, it is better to make them fairly simple. If possible, each student should have one Gram-positive and one Gram-negative organism. You may or may not want to give the students a list of organisms present in the mixes.
2. If you include β *Streptococcus* in your unknown mixes, it is important to have blood agar plates, otherwise TS agar should work very well for the complete medium.
3. A MacConkey agar plate is used here because it is commonly used in clinical laboratories to isolate Gram-negative organisms. It contains lactose, a pH indicator, and dyes to inhibit Gram-positive organisms. Be sure to use MacConkey with crystal violet because the color is darker for lactose positive organisms and the dye *somewhat* reduces the swarming of *Proteus*—especially if the plates are dry. MacConkey agar without crystal violet will support the growth of some Gram-positive enterococcus, which is desirable in some clinical situations.
4. You would expect organisms appearing purple on MacConkey (and therefore lactose positive) to ferment lactose in the lactose fermentation tube.
5. When preparing mixes of organisms, grow separately overnight in TS broth. When mixing together, you usually must add more Gram-positive organisms relative to the Gram negative so that both organisms will appear in about the same number on a streak plate. Some examples:
   One part *E. coli* to 3 parts *Staphylococcus epidermidis*
   One part *E. coli* to 4 parts *Bacillus*

6. Provide examples of positive tests when students are observing their results.

Exercise 27

Students should be made aware that all human blood is potentially infectious, even though it is screened by the blood bank before use for eight infectious agents, including AIDS and hepatitis (A, B, and C) viruses. This is one reason a dropping bottle is preferred to other means (e.g., a capillary pipet) for dispensing a drop of blood. Safety glasses and disposable rubber gloves should also be made available to students so inclined.

Exercise 29

Proteus OX-2 antigen for slide test, 5 ml, Difco Cat. #2243-56-6
Proteus OX-2 antigen for tube test, 25 ml, Difco Cat. #2248-65-0
Proteus OX-2 antiserum, for slide or tube test, 3 ml, Difco Cat. #2245-47-6
A recommended source for *Proteus* OX-19 antigen and positive polyvalent control antiserum is Difco Laboratories. Difco has an educational program whereby their short-dated products are provided free to educational institutions. Contact your local Difco representative about availability. *Proteus* OX-19 antigen is available in either 5-ml (slide test, Difco Cat. #2234-56-7) or 25-ml (tube test, Difco Cat. # 2247-65-1) quantities. The positive polyvalent control serum is available in 5-ml quantities (Difco Cat. # 3238-56-1). It should be mentioned that the 5-ml quantity of OX-19 antigen can't be used for the tube test. *Suggestion:* Use of *Salmonella* H antigen (a,b, or d) instead of *Proteus* OX-19 antigen in that it can be used interchangeably with both the slide and tube test. The same positive polyvalent control serum is used for both the slide and tube test.

The USR cardiolipin syphilis antigen (six 3-ml vials, Difco Cat. #2405-46-3) and USR positive control serum set (3 vials: positive, weakly reactive, and negative, Difco Cat. # 3516-32-3) are also included in Difco's short-dated products.

For the Weil-Felix test it is possible to dilute the positive polyvalent control antiserum before use in that it has a very high titer (1:80 or greater) as determined by Difco Laboratories. The significant titer for *Proteus* OX-19 agglutination with the positive polyvalent control antiserum is 1:160–1:320. The significant titer for *Salmonella* H antigen d with the same antiserum is 1:160 or higher. By definition the polyvalent positive control antiserum is active against a host of "febrile antigens." "Febrile antigens" refers to a number of species of microorganisms pathogenic to man and which are accompanied by a fever in the host. The most prominent organisms in this group include the genera *Salmonella, Brucella, Francisella (Pasteurella), Leptospirea,* and some of the *Rickettsia.* Later, Weil and Felix discovered a fortuitous cross-reaction between the agglutinins produced by the causative agent of typhus fever, *Rickettsia prowazekii,* and certain strains of *Proteus vulgaris* isolated from patients suffering from typhus fever. The *Proteus* strains designated OX-19 and OX-2 seem to share an alkali-stable polysaccharide with *R. prowazekii* and therefore produce agglutinins identical to it.

You may wonder why the rapid slide test, a qualitative test for serum antibodies, employs five serum dilutions. The five dilutions are necessary to detect the agglutinin content of a serum which might be overlooked, due to sometimes producing a negative serum agglutination reaction with a higher serum agglutinin concentration. The latter is particularly true with *Brucella* and to some extent with typhoid agglutinins.

Exercise 30

The BBL Strep Grouping Kit (Catalog #4340860) may be obtained by contacting either your local BBL representative or laboratory supplier. A discount price is usually offered when the kits are used for educational purposes. Occasionally outdated but still usable kits are available. To obtain the latter you must write a letter stating that the kit is for educational use and not for clinical diagnostic use. The letter should be addressed to:
Technical Services Department
Becton Dickinson Microbiology Systems
250 Schilling Circle
Cockeysville, MD 21030
(Tel: 1-800-638-8663)

Each kit contains two vials of lyophilized cell wall extraction enzyme. To reconstitute, transfer the contents of a vial into a clean 25- or 50-ml

Erlenmeyer flask, and dissolve with 12 ml of sterile, purified (deionized) water. Pipette 0.3 ml aliquots of the diluted extraction enzyme into clean serological test tubes. Each student or group of students will require 3 tubes of extraction enzyme.

Exercise 31

Abbott Laboratories has kindly consented to provide, for educational use, complimentary TestPack Plus Strep A ELISA test kits within their budgetary limitations. Please forward your request for the above described kit to:
Ann G. Shaha
Senior Product Manager
Abbott Laboratories, 2KC, Bldg. AP6C
Abbott Park, IL 60064

Exercise 32

Meridian Diagnostics, Inc., has kindly agreed to provide, for educational use, usable outdated *Candida* immunodiffusion test kits at a reasonable cost. Please direct your inquiries to:
Manager of Technical Services
Meridian Diagnostics, Inc.
3471 River Hills Drive
Cincinnati, OH 45244
(Tel: 1-800-343-3858; in Ohio call collect: 513-271-3700)

The following information pertains to materials used in the test procedure.

The immunodiffusion agar plates contain 0.9% agarose in glycine buffer at pH 7.4 ± 0.2.

The antigen used is *Candida albicans* strain B385, available from Meridian Diagnostics. This strain was chosen for antigen production because it demonstrates antigens in common with all species of *Candida* except *C. krusei* (Stickle, D. 1972. *Appl. Microbiol.* 23:490–499).

Antibodies for positive control serum are produced in either immunized goats or rabbits. Properly produced serum will produce 2–3 precipitin bands against *Candida* antigen. The serum is preserved with sodium azide, a potentially explosive chemical. For this reason it is recommended that excess control serum be discarded in a hazardous waste container.

The volume of water required to rehydrate the control serum is dependent on the amount of dehydrated control serum; e.g., if the dehydrated label volume reads 0.4 ml, then rehydrate with 0.38 ml of reagent-grade water. If the dehydrated label volume reads 1.0 ml, rehydrate with 0.95 ml of reagent-grade water.

Mix by gently flicking the bottom of the vial with your fingertip.

Allow the vial to stand at room temperature for 1 hour; invert occasionally to mix **(do not shake).**

If upon first use, the serum appears inactive, it may be that sufficient standing time was not allowed. If so, repeat the test.

Any control serum which will not be consumed within one month after rehydration should be aliquotted and stored in the freezer. At $-20°$ C a minimum 9-month life is possible. Repeated freezing and thawing should be avoided. Also, when in use, the time which the control serum is kept at room temperature should be as short as possible. Rehydration requires 1 hour and serum activity should be established beforehand.

Exercise 33

This exercise requires four working sessions for its completion. But it is not as complicated and time-consuming as it seems. The exercise has been very educational for our students and seems worth the time involved. It emphasizes how the students can be agents of infection transmission, why infection control and isolation procedures in the hospital are important, and what can be learned from data obtained by epidemiological methods. Student time can be minimized by scheduling in the following ways: (1) This exercise will only take a few minutes for most sessions, so it could be scheduled along with other exercises. (2) Alternatively, if students are permitted to work independently outside regular class hours, they can come in on their own for subculturing; they will have already learned these procedures in previous exercises. (3) Not all students will have positive cultures so some sharing of the work would be possible.

Some strains of *S. epidermidis* seem to be able to ferment mannitol, so this should be carefully considered in the screening procedure.

In the past the reference strain has been susceptible to this battery of antibiotics, but occasionally has failed to produce beta hemolysis. However, it seems advisable to classify as resistant any strain which shows resistance to any of the antibiotics, regardless of the reference antibiogram. We found quite a few (13 out of 54 strains) coagulase-negative strains; this may be student error, so these will be checked more carefully next time. We considered positive *S. aureus* as any mannitol fermenter showing hemolysis or coagulase production.

The format of the Information Sheet makes tabulation very simple. This can be done by dividing the slips into two stacks, hospital workers and nonworkers. Each group is sorted again according to positive carriers and noncarriers, and each group counted. The carrier groups can then be sorted into resistant and susceptible strains, and counted. This process can be repeated until all categories have been counted and recorded.

Exercise 34

Gelman Sciences, 600 South Wagner Road, Ann Arbor, MI (Tel: 1-800-521-1520) recently made available a disposable membrane filter apparatus which is easy to use and inexpensive. Following filtration, it can be converted into a Petri dish for incubation of the membrane. Its design is such that a sterile 2-ml ampoule of Endo liquid medium can be added aseptically from below to the underside of the filter pad. The approximate cost per unit including the Endo medium is $4.50. You will also need a side-arm flask and a #8 rubber stopper for vacuum filtration. Call Gelman for an illustrated brochure.

INDEX

Neisseria, 208
Nemathelminthes, 180
Neutrophil, 231, 232
Noncoliform bacteria, 301
Nonionizing wavelength of radiation, 97, 98
Nonspecific immunity, 229
Nosocomial infection, 287
Numerical aperature, 17

O

Objective lens, 13
Obligate aerobe, 73
Obligate anaerobe, 73
Oil-immersion objective, 12
Operon, 145
Opportunistic infection, 153, 160, 170, 274, 275
Optical density (O.D.), 63, 64, 77, 78
Osmotic pressure, 105–6
Ouchterlony double immunodiffusion test, 273–75
Oxidase, 208
Oxidative pathway, 160
Oxygen, microbial growth and, 73
Oxyuris, 181

P

Para-aminobenzoic acid (PABA), 113, 114
Paracoccidiodes braziliensis, 159
Paramecium, 178
Parasitic diseases, 153–54, 177–82
Parfocal objective, 17, 33
Pasteur, Louis, 85
Pasteurization, 87, 88
Pathogenic organisms, 3, 52, 88, 116, 158, 159. *See also* epidemiology; medical microbiology
Penicillin, 153
Penicillium chrysogenum, 153
Peptidoglycan, 40
Peritrichous flagella, 45
Phagocytic cell, 229
Phenol, 114
Photosynthetic bacteria, 3
Phycomycetes, 169
Pinworms, 154, 180–81
Planarians, 179, 181
Plaque, viral, 154–55, 190, 191
Plasma, 232
Plasmid, 137–38
Plasmodium, 178
Plasmolysis, 105, 106

Plasmoptysis, 105, 106
Plate count, 63, 64, 301
Polar flagella, 45
Pollution, of water, 299
Precipitin reaction, 229, 256, 273, 275
Proglottid, 179, 181
Prokaryote, 153, 189–91
Propionibacterium acnes, 201
Propionibacterium granulosum, 201
Proteolytic enzyme, 237
Proteus, 213, 214, 243–44
Prototroph, 123
Protozoa, 177–82
Pseudohyphae, 157, 160
Pseudomonas, 73
Pseudopodia, 178, 181
Public health, 286, 299–304
Pyriform, 170
Pyrimidine dimer, 97, 98

Q

Quantification, of microorganisms, 63–64

R

Ramsden's disk, 13, 17
Rapid plasma reagin (RPR) 18 mm circle card test, 244
Reagin, 244, 245
Red blood cells, 231–32
Red tide, 299
Refractive index, 15, 17
Resolution, microscope, 14, 17
Respiratory microorganisms, 207–8
Resting stage cell, 23
Rheumatic fever, 207, 255
Rhizoids, 170
Rich medium, 59
Rickettsial diseases, 243
Rickettsia rickettsii, 243
Ringworm, 168, 170
Rocky Mountain spotted fever, 243

S

Saccharomyces cerevisiae, 157, 158
Saccharomyces spp., 153
Saccharophilic microbes, 106
Salmonella, 213, 243, 303
Salt concentration, 105–6
Sarcodina, 178
Schistosoma mansoni, 184
Schistosoma spp., 179
Schistosomiasis, 184
Schizont, 181

Scolex, 179, 181
Selective medium, 51, 59, 219, 287, 301, 304
Selenotila intestinalis, 157
Semipermeable membrane, 105, 106
Sequala diseases, 255, 256
Serial dilution, 64, 190, 191
Serodiagnosis, 243, 245, 275
Serological test, 208
Shigella, 213
Simple microscope, 17
Simple stain, 33
Sleeping sickness, 178
Smallpox, 286
Smear, 33
Solute concentration, 105
Somatic antigen, 245
Specific immunity, 229
Spectrophotometer, 63, 77, 78
Sphaerotilus natans, 299, 300
Spherical lens aberration, 12, 17
Spherule, 170
Spirillum, 43
Spontaneous generation, 85
Sporangiospore, 170
Sporocyst, 181, 184
Sporotrichum schenkii, 169
Sporozoa, 178–79
Staining, of microorganisms, 31, 33
Staphylococcus aureus, 201, 287–88, 303
Staphylococcus epidermis, 201
Staphylococcus spp., 59, 105–6, 208
Star diaphragm, 23
Sterigma, 171
Sterile, 5, 53
Sterilization, heat, 87–88
Storage granules, 43, 44
Streak-plate culture, 51, 55–56
Strep throat, 207
Streptococcus faecalis, 301
Streptococcus mutans, 43
Streptococcus pneumoniae, 207, 208, 255
Streptococcus pyogenes, 207, 255
Streptococcus spp., 33, 255–57, 263–64, 265, 267–68
Streptomycin, 125, 131
Subcutaneous mycoses, 168–69
Substrate, 145
Sulfanilamide (sulfa drug), 113, 114
Surfactant, 301, 304
Syphilis, 237, 244

T

Taenia saginata, 180, 181
Tapeworms, 177, 179–80
Temperate phage, 190, 191

Temperature, microbial growth and, 5, 77, 80
Thermal death point (TDP), 87–88
Thermal death time (TDT), 87–88
Thermoduric, 87, 88
Thermophilic, 87, 88
Throat culture, 207–8
Thymidine, 97, 98
Titer, 189–91, 245
Tobacco Mosaic Virus (TMV), 154
Toxic Shock Syndrome, 201
Toxoplasma, 179
Transcription, genetic, 145
Transduction, genetic, 123
Transformation, DNA-mediated, 123, 131
Trematoda, 179
Treponema pallidum, 244
Trichinella spiralis, 180
Trichomonas vaginalis, 154
Trigonopsis variablis, 157
Trophozoites, 178, 181
Trypanosoma gambiense, 178
Trypticase soy agar, 59
Tubellaria, 179
Tuberculosis, 41
Turbidity, 63, 64
Turbidometric count, 63–64

Tyndall, John, 85
Tyndallization, 85, 87, 88
Typhoid fever, 299

U

Ubiquity, 5
Ultraviolet (UV) light, 97–98
Undefined medium, 59, 60
Unheated serum reagin (USR) test, 244
Urea hydrolysis test, 214

V

Vaginitis, 154
Van Leeuwenhoek, Anton, 11–12
Vector, 243, 245
Vegetative cell, 40
Venereal Disease Research Laboratory (VDRL) test, 244
Virology, 154
Virulent phage, 190, 191
Viruses, 154–55, 189–91, 207
Visible spectrum, of light, 98
Vital stain, 24
Voges-Proskauer test, 214

W

Wasserman complement fixation test, 237
Water, bacteriological examination of, 299–304, 308
Weil-Felix test, 243–44
Wet mount slide, 17, 23, 24
White blood cells, 229
Widal test, 243
Wild type, 123, 126
Working distance, 15, 17
World Health Organization (WHO), 286
Wright's stain, 231

Y

Yeasts, 23–24, 105, 106, 153, 157–60, 208

Z

Zone diameter, 115